WAREHOUSE CONVEISION

仓库办公室

摄影：小白 撰文：钟音

辽宁科学技术出版社
LIAONING SCIENCE AND TECHNOLOGY PRESS

目　　录

CONTENTS

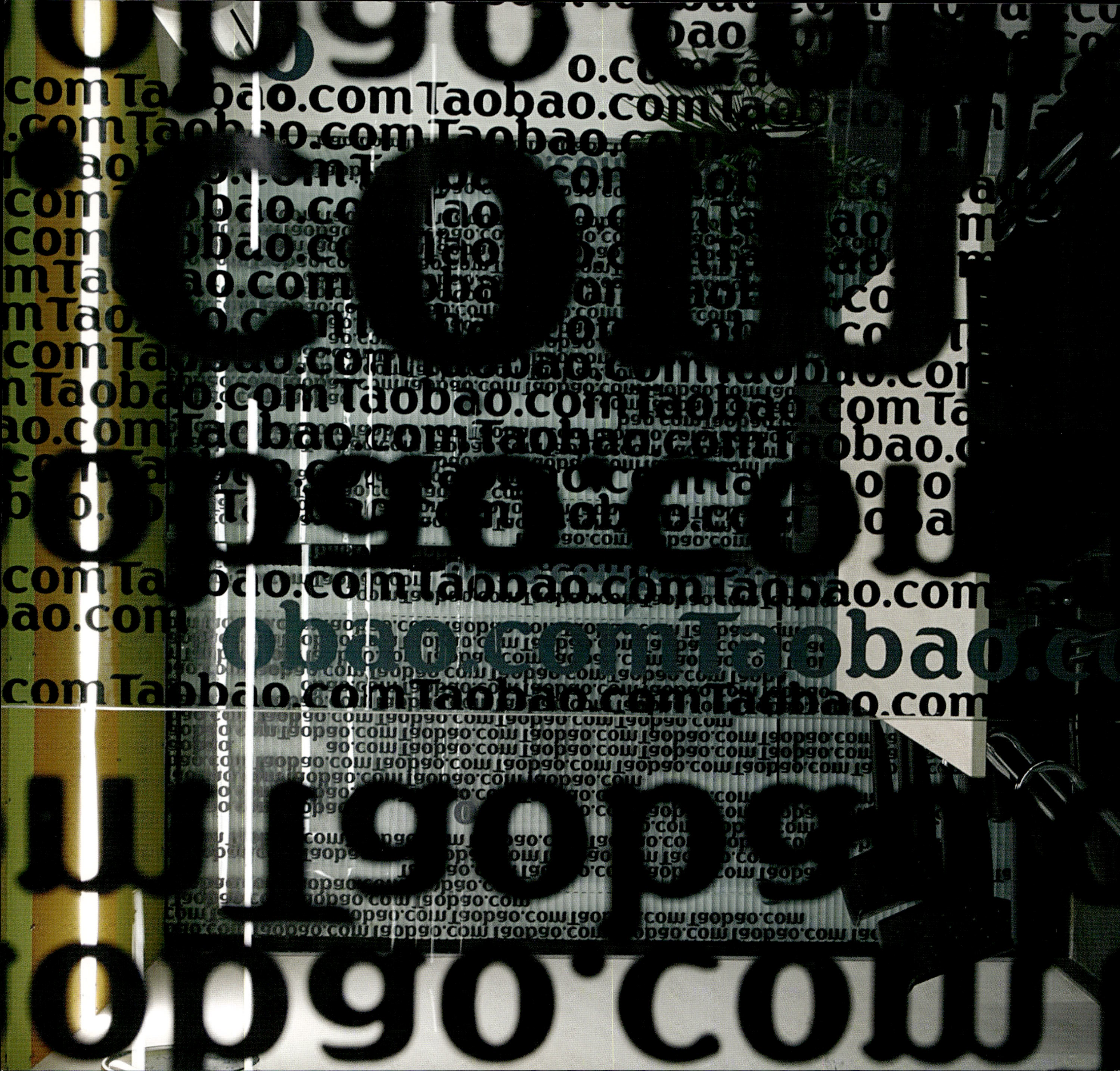
Taobao.com

前　言 PREFACE

建筑作为城市给人的第一印象，成为人们的比较对象和乐此不疲的话题。随着时代的发展，建筑承载了更多内涵，建筑带来了生活方式的改变、观念的进步，建筑也构成了都市特有的气质。

一个飞速发展的城市，每时每刻都在以无法比拟的速度向前发展着。而在这发展进程中，由于产业结构升级与城市功能转换的因素，造成了众多老厂房、老仓库、旧建筑逐渐空置或废弃。本书的设计师们运用新的设计理念，保留了老厂房特有的建筑风格和历史底蕴，为历史的留存注入时尚、创意的元素，将这些荒废的旧仓库脱胎换骨为一幢幢拥有独具特色的办公场地。

在这个充满喧嚣的城市里，设计师将许多或矛盾或协调的设计元素错综复杂地交缠在一起，融入到这些充满了艺术、充满了现代时尚感的办公楼中，为现代人带来了更多创意的灵感，更多灵感的源泉。这些建筑也改善了城市环境，提升了城市功能，成为了现代城市景观的一道新景象。

As the first impression of a city, architecture becomes a topic that people would like to talk about and an object that people would like to compare with. During the course of development of the society, architecture bears more meanings, it brings about the change of lifestyle, advancement of ideas and etc, and meanwhile it contributes to the identity of a city.

A rapidly developing city goes forward without any cease in an incomparably flying speed. Nevertheless, during the course of its development, many old plants, workshops, warehouses and structures were not in use or abandoned due to the factors such as update of industrial structure and conversion of urban function. In view of these, the author of this book exert novel designing ideas to reserve the owned architectural styles and historical connotations of many old plants and workshops, to infuse voguish and originative elements into them and to convert these dilapidated old structures into various featured office buildings.

Out of this blatant city, the designer makes an anfractuous mix of many contradictory or otherwise harmonious designing elements and then integrates it into these artistic and modern voguish office buildings which become a fresh vision in the modern city and bring more originative inspirations and more sources of inspirations, and as well as improve the environment of the city and exalt the function of the city.

Forte Fucheng
复城国际

位置：上海市曲阳路910号 电话：+86－021－65530055

复城国际拥有一个相当大气的办公空间，超高的挑空设计和丰富多样的视觉层次给工作室带来了更大的创造空间。那些明快的色彩、简洁的构成以及对空间的良好感觉，在这个艺术空间里得到了丰富的体现。

设计师充分发挥了复城国际开阔大厅的优势，在内部结构上采用了跃层设计，将办公环境按商务、办公、休息等不同功能划分为互不干扰的立体空间，使原本平面单调的办公楼展现出自由灵动的氛围。内墙是直接裸露的红色糙砖，辅以凌乱的白石灰，颇具艺术气息，与古朴的淡黄色甚至有些裂痕的粗糙外墙风格近似。大厅中间是个高出地面约半尺的矩形玻璃平台，一侧连接着一个通往二楼的楼梯，楼梯的踏板采用同样的玻璃材质，在扶手及接地处都镶嵌着银色的金属条，在黄色顶灯的照射下，折射着金色的光芒，极富现代时尚感。斜穿大厅的是空中一条长长的廊道，本设计采用深色木板的走道、带着锈斑的篱笆状护栏，与光滑剔透的玻璃扶栏，使楼梯显得不再沉闷与古板。

大厅左侧摆放着许多休闲沙发，开放式的设计令空间更加自由自在。右边则是VIP空间，用木帘制成的半隔断设计既保留了视觉上的延展性，又强调了办公区域的私密性。流线型的太空金属桌面也是复城国际的一大亮点，流畅的线条活跃了整个空间，别出心裁的流动设计更是让每一个人的心情也随之跳跃起来。

Forte Fucheng has a very huge office space, its superbly high-rising space and diversiform visual levels bring forth bigger creative space to the studio. Those bright colors, simplified forms and perfect sense of space are all expressed fully.

The designer fully exert the super openness of the hall of Forte Fucheng, running-over design is adopted interiorly, office space is divided into non-interruptive minor spaces according to different functions commercial, official, resting and etc. which make the premier monotonous office building clever and free. The wall interior is made of red coarse bricks, randomly finished by white lime. The style of the wall interior is similar to that of some light yellowish old cracked exterior wall and seems full of artistic flavor. At the middle of the hall is a rectangular glass platform that is half meter higher than the level of the ground. One end of the platform is connected to a staircase leading towards the second floor, the treads is the same with platform in material. Metal strips in silver color are decorated on where the balustrades touch the floor. Under the shining of the yellowish skylight, the strips reflect golden shines, which is extremely full of sense of modern vogue. A long corridor runs diagnostically through the hall, the design is to emphasize the dark colored wooden-planked way, rust spotted fence-like balustrade and smooth transparent glass rails which weaken the monotony and stiffness of the staircase.

At the left part of hall are some leisure sofas. The open styled design makes the space more free and comfortable. VIP space is on the right; wooden half partition wall made both reserves the visual prolongation and emphasizes the privacy of the office area. Streamline table surface with effect of Space metal is a highlight of Forte Fucheng, the fluent lines vivify the entire space, and fantastic fluid design stimulates everybody's mood.

P6-P7

1．流畅的线条活跃了整个空间。

1. Fluent lines activate the entire space.

1．木制的田园气息与金属的现代气息形成了强烈的对比。
2．别出心裁的流动设计更是让每一个人的心情也随之跳跃起来。
3．腾空的走廊下是会客区，开放式的设计令空间更加自由自在。

1. A strong contrast occurs between garden taste formed by wood and modern taste formed by metal.
2. Fantastic fluid design stimulates everybody's mood.
3. Parlor area is below the vacant corridor and open styled design makes the space more free and comfortable.

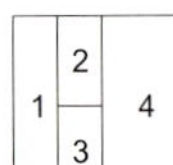

1. 独具艺术气息的文化墙。
2. 半隔断的设计既保留了视觉上的延展性，又强调了办公区域的私密性。
3. 办公区一角。
4. 超高的挑空设计和丰富多样的视觉层次给工作室带来了更大的创造空间。

1. Cultural wall full of artistic taste.
2. Half partition made of wooden curtain both reserves the visual prolongation and emphasizes the privacy of the office area.
3. A part of the office area.
4. Superbly high-rising space and diversiform visual levels bring forth bigger creative space to the studio.

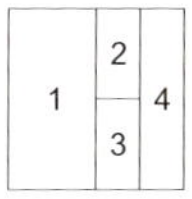

1．本设计采用深色木板的走道、带着锈斑的篱笆状护栏，与光滑剔透的玻璃扶栏，使楼梯显得不再沉闷与古板。
2．玻璃的材质强调了通透性。
3．楼梯底部的白色灯光透过蓝绿色的玻璃，幽幽地散发着淡淡的光晕。
4．银色的金属条在黄色顶灯的照射下，折射着金色的光芒。

1. The design is to emphasize the dark colored wooden-planked way, rust spotted fence-like balustrade and smooth transparent glass rails which weaken the monotony and stiffness of the staircase.
2. Glass material emphasizes transparency.
3. White lamplight through blue-greenish glass at the bottom of the staircase gives off tender shines.
4. The silver colored metal strips reflect golden shines under the lighting of the yellowish overhead light.

Mi Qiu Studio
米丘现代艺术工作室

位置：上海市东大名路713号6楼　电话：+86－021－62131352

米丘工作室是一个典型的由仓库改建而成的工作室。整个设计风格强调现代人对空间的渴望，讲究无论室内室外都想要有的透气感。也是现代人追求自我、追求自由的体现。

米丘工作室的空间十分开阔，宽敞的空间内几乎找不到一扇门。入口处高大的飞人雕塑象征着自由翱翔的创意灵感。设计师充分利用大幅落地玻璃窗和光滑的金属隔离墙，并提供大量共享空间，以提高自然光线的通透性。工作室保留了原来的墙面和房顶，带着斑驳的墙体和立柱自然而又古朴，浅灰色的房顶上横竖交错的管道与房梁延伸了空间，工作室内还到处摆放着年代各异的古董家具和造型各异的雕塑。这些传统的设计元素与充满现代感的玻璃和金属交缠在一起，体现了传统中的时尚。

整个工作空间的组合也非常富有创意。中间是一个由大幅玻璃构成的会议室，落地的透明玻璃，没有距离，无拘无束，使会议室与整个空间融为一体。在工作室的一侧，是开放式的工作空间，空间开敞而贴近自然。在另一侧，设计师设计了一堵直角的金属隔离墙，一面的金属墙以其光滑的质地在透着微光、晶莹通透的蓝色灯箱下，营造了光与材质的虚幻，使空间更富于变化。还有一面金属墙的背面则被设计成了一个“顶天立地”的书橱，自由的空间既通透流畅，又体现了工作室的功能。

Mi Qiu Studio is a workroom typically rebuilt from a warehouse. The whole style emphasizes the modern people's desire to the space and sense of free breath both on interior and exterior. This is also an embodiment of pursuing ego and freedom.

Mi Qiu Studio is spacious with no door can be found. The sculpture flying man at the entrance symbolizes free hovering inspiration of originality. The designer takes full use of big clerestory and smooth metal partition and provides much shared space to improve transfixion of natural light. The studio reserves the original wall and roof, stained wall and pilasters are natural and primitively simple, interlaced ducts and girder on light grayish roof extend the space. Many pieces of curio furniture in different ages and statues in various forms are displayed around the interior of the studio. These traditional design elements are tangled with modern glass and metal, which presents the traditional vogue.

The combination of the entire working space is very originative. The middle is a meeting chamber made of big pieces of transparent glass. As the pieces of glass fall to the ground, it seems no distant and unconstraint and the chamber is in harmony with the entire space. At the side of the workroom is the opening working area, and the space is spacious and very close to nature. At the other side of the workroom, the designer sets a metal partition wall of right angle. Metal wall on one side reflects tender light through its smooth surface material. The illusion of light and material occurs under the crystal and bluish cased lamp, which makes the space transformative. A giant bookcase is set against the back of a piece of metal wall, which makes the space free and fluent and meanwhile the function of studio is obvious.

1. 工作室保留了原来的墙面和房顶，带着斑驳的墙体和立柱自然而又古朴，浅灰色的房顶上横竖交错的管道与房梁延伸了空间。

1. The studio reserves the original wall and roof, stained wall and pilasters are natural and primitively simple, interlaced ducts and girder on light grayish roof extend the space.

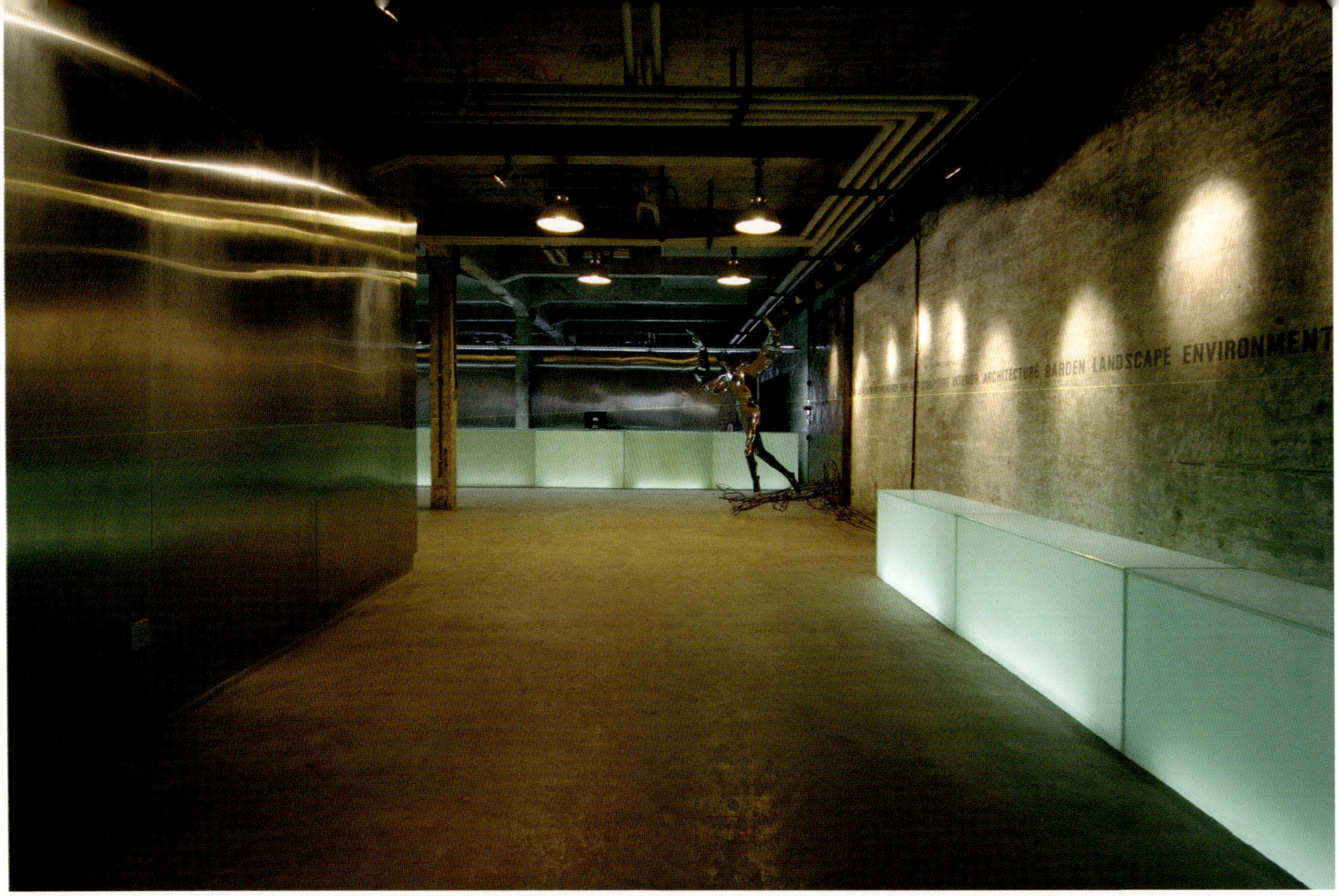
ARCHITECTURE GARDEN LANDSCAPE ENVIRONMENT

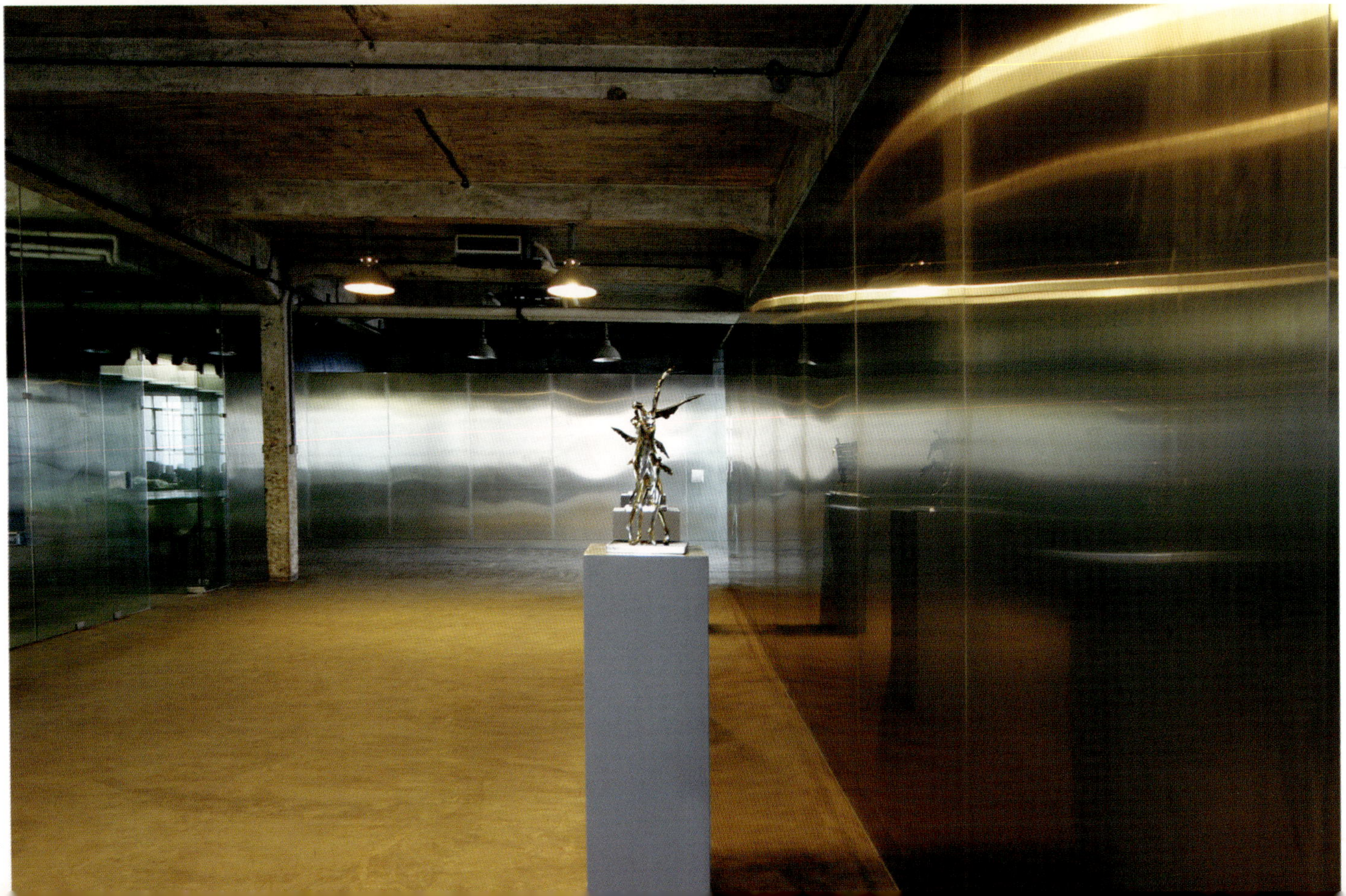

1 3
2

1．金属墙以其光滑的质地在透着微光、晶莹通透的蓝色灯箱下，营造了光与材质的虚幻，使空间更富于变化。
2．在工作室的另一侧，设计师设计了一堵直角的金属隔离墙。
3．落地的透明玻璃，没有距离，无拘无束，使会议室与整个空间融为一体。

1. Metal wall on one side reflects tender light through its smooth surface material.
2. At the other side of the workroom, the designer sets a metal partition wall of right angle.
3. Falling to the ground, the pieces of transparent glass seem no distant and unconstraint and the chamber is in harmony with the entire space.

P18-P19

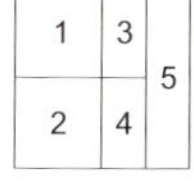

1．整个设计风格强调现代人对空间的渴望，讲究无论室内室外都想要有的透气感。
2．金属墙的背面则被设计成了一个“顶天立地”的书橱，自由的空间既通透流畅，又体现了工作室的功能。
3．入口处高大的飞人雕塑象征着自由翱翔的创意灵感。
4．开放式的工作空间，空间开敞而贴近自然。
5．工作室内到处摆放着年代各异的古董家具和造型各异的雕塑。

1. The whole style emphasizes the modern people's desire to the space and sense of free breath both on interior and exterior.
2. A giant bookcase is set against the back of a piece of metal wall, which makes the space free and fluent and meanwhile the function of studio is obvious.
3. The sculpture flying man at the entrance symbolizes free hovering inspiration of originality.
4. The opening working space is spacious and very close to nature.
5. Many pieces of curio furniture in different ages and statues in various forms are displayed around the interior of the studio.

MI QIU MODER

MADA s.p.a.m.
马达斯班

位置：上海市新乐路134弄2号 电话：+86－021－54041166

马达斯班的整体结构采用了跃层设计。整个MADA S.P.A.M OFFICE始终围绕白、灰、浅褐这三大色彩元素，给人感觉既协调又整洁。

入口楼梯的设计让人眼前一亮，加长加宽的灰色台阶上镶嵌着浅褐色的仿马赛克地砖，灰色与褐色有规则地层叠向上，不仅增强了延伸性，更让人有一种艺术的享受。最让人叫绝的是入口处的楼梯没有扶手，让人在视觉上感觉自由自在，无拘无束。

沿着蜿蜒曲折的楼梯向上，踩着仿马赛克的台阶，仿佛使人置身于朴实无华的颜色与绝佳天然质感所构筑的空间中。四周是挑高的空间，大幅的落地玻璃墙与条形的透明玻璃窗，强调了采光的重要性，在视觉上令空间显得更加通透明亮。二楼的工作区无论是墙面还是地面都大量采用了仿马赛克的材质，不仅质朴厚重，而且细腻润泽，颇具深厚的文化内涵。色彩丰富的褐色小块与白色的房顶对比，增加了立体感，也丰富了工作室的空间。简单的白色也正因为有了褐色小块的映衬，也跳出了单调的白色，整个OFFICE给人一种典雅的感觉。

为了倡导贴近自然的意境，设计师在办公区半围合的区域独具匠心地设计了玻璃顶棚，充足的自然光透过大块的透明玻璃，营造了明亮、安静、舒适的工作环境。玻璃通过自身材料的特性拉近了人与环境的距离，模糊了空间与自然的概念，改变了办公楼给人们带来的压抑感，增加了空间情趣。

The overall structure of MADA S.P.A.M. adopts over-floor design. MADA S.P.A.M. is dominated by three colors white, grey and light brown, which give people a sense of harmony and tidiness.

The staircase at entrance attracts people's eyes very much. Lengthened and widened grayish steps are decorated with light brownish mimic mosaic tiles. Grayness and brownness cascade upwards regularly, which builds up the sense of prolongation and enjoyment of artistic appeal. The most attractive is that there are no handrails on the staircase at entrance area. It makes people feel free and unrestrictive visually.

If one treads on the mimic mosaic tiled steps and go along the winding staircase upward, he will find himself in the space structured by plain colors and perfect natural materials. Heightened spaces are around. Huge pieces of glass walls which touch the ground emphasize the importance of lighting effect and make the space more bright and visualized. At working area on the second floor, both the walls and ground adopt more mimic mosaic materials which are not only rustic and massy but also slender and smooth. The materials contain profound cultural implication. The contrast between lush brown pats and white ceiling both increases the sense of cupidity and enriches the space of working workroom. Simple white is no longer monotonous because of the presence of small pieces of brown pats. The entire office conveys people a sense of elegance.

In order to inspire the artistic conception of approaching to nature, the designers designed glass ceiling at the semi-enclosing area within working area. Abundant sunshine travels through the huge pieces of glass, which contributes to a bight tranquil and comfortable working environment. Glass serves as a media between human being and the environment through its own property. It blurs the boundary between space and nature, changes the usual sense of oppression which former office buildings bring forth to people and adds spice to the space.

1．灰色的台阶上镶嵌的浅褐色仿马赛克地砖是画龙点睛之笔，增强了延伸性，也让人有一种艺术的享受。

1. Lengthened and widened grayish steps are decorated with light brownish mimic mosaic tiles. Grayness and brownness cascade upwards regularly, which builds up the sense of prolongation and enjoyment of artistic appeal.

P22-P23

1	3
2	

1．色彩丰富的褐色小块与白色的房顶对比，增加了立体感，也丰富了工作室的空间。
2．高挑的空间也是这一工作室设计的一大亮点。
3．大幅的落地玻璃墙与条形的透明玻璃窗，强调了采光的重要性，在视觉上令空间显得更加通透明亮。

1. The contrast between lush brown pats and white ceiling both increases the sense of cupidity and enriches the space of working workroom.
2. Heightened spaces design is a highlight of this office.
3. Huge pieces of glass walls which touch the ground emphasize the importance of lighting effect and make the space more bright and visualized.

1		3
	2	

1．充足的自然光透过大块的透明玻璃，营造了明亮、安静、舒适的工作环境。
2．工作室始终围绕白、灰、浅褐这三大色彩元素，给人感觉既协调又整洁。
3．走在仿马赛克的楼梯上，仿佛使人置身于朴实无华的颜色与绝佳天然质感所构筑的空间中。

1. Abundant sunshine travels through the huge pieces of glass, which contributes to a bight tranquil and comfortable working environment.
2. The studio is dominated by three colors white, grey and light brown, which give people a sense of harmony and tidiness.
3. Treading on the mimic mosaic tiled steps and going along the winding staircase upward makes one feel himself in the space structured by plain colors and perfect natural materials.

Sihan Design
思瀚设计

位置：杭州市文二路391号3号楼中跨3层 电话：+86－0571－88480363

石库门最能充分展现上海里弄的特有风情，同样，在杭州以石库门建筑形式作为主要设计元素的思瀚设计工作室自然就会受到越来越多人的关注。思瀚的整个空间以传统风格和现代实用为依托，在细节方面也独具特色;精美的雕刻、古老的装饰物、凹凸有致的门楣，处处显露着形似单纯却折射着东方特有的稳重沉着，让整个空间沉浸于东方情结洗礼中。

入口处是风情万种的石库门弄堂，门口的磨具、青竹、砂石，厚重的大门，卷涡状的门楣，老旧的砖石，还有墙上的斑纹，皆在体现古朴的传统。

室内的设计同样也延续了这样的元素。深浅交错的灰色砖墙，斑驳门梁上精美绝伦的砖雕，形状各异的古老装饰物，都在空间诠释着东方的妩媚。

思瀚有好几间独立的工作室，通往这些工作室的过道上都采用了错层的设计，弥补了平面空间带来的平淡感，使空间变得更丰富更灵动。其余过道上的工作室全部由落地玻璃墙分割而成。特别值得一提的是思瀚设计工作室的地砖也各不相同。青砖步行道让人仿佛置身于石库门弄堂。而浅褐色地砖上则描绘着大幅的莲花图，有迎头傲放的，也有含苞欲放的，都以其独特的姿态在空间中娇艳地展示着令人无法抵挡的东方诱人魅力。

Stone gates can best show the special taste of lanes in Shanghai. Similarly, regarding stone gate as main design element let Sihan Design focused by more and more people. The entire space of Sihan is based on traditional style and modern practicality. Characteristics are in every detail: subtle engravings, antique ornaments, concavo-convex lintels all show the simplicity yet depth and steadiness specially owned by the orient. The entire space is immersed in the oriental complex.

At entrance are alley full of tastes, grind tool, bamboo, gravel stones, bulky doors, lintel of scroll form, old bricks and streaks on the wall, each of which embodies primitive and simple tradition.

Interior follows the same design elements. Interlaced dark and light gray bricked wall, subtle brick-engravings on rustic architrave and antique ornaments all explain oriental charms within the space.

Sihan Design has several independent studios, aisles leading to which adopt alternate floor design which offsets the plainness brought by plane space and makes the space clever and flexible. The studios at the sides of the rest aisles are all divided by glass partition walls. What is worth mention is that the floor tiles in the studios are different each other. Pedestrian path of black brick makes the people find themselves as if in real stone gate alley. The lotus patterns on the light brown floor tiles are various: rampant blooming, ready to bloom in bud and etc. no matter in what pattern, they show irresistible oriental charms in the space respectively in unique way.

P26-P27

	1

1．厚重的大门，卷涡状的门楣，老旧的砖石，还有墙上的斑纹，仿佛都在诉说着石库门的故事。

1. Bulky doors, lintel of scroll form, old bricks, and streaks on the wall all as if tell the story of the stone gate.

1	2

1．浅褐色的地砖上描绘着大幅的莲花图，有迎头傲放的，也有含苞欲放的，都以其独特的姿态在空间中娇艳地展示着令人无法抵挡的东方诱人魅力。
2．门梁上的砖雕，精美绝伦，为整个工作室增添了几分庄严、优雅与个性。

1. The lotus patterns on the light brown floor tiles are various: rampant blooming, ready to bloom in bud and etc. no matter in what pattern, they show irresistible oriental charms in the space respectively in unique way.
2. The brick-engravings on the architrave are perfectly beautiful and they add grandeur, elegance and individuality to the entire studio.

1．弧形的吊顶带来了空间的流动感。
2．怒放的莲花与落地的玻璃墙新旧对话，交相辉映。
3．青砖步行道让人仿佛置身于石库门弄堂。
4．小小的装饰物也为工作室增色不少。

1. Arc suspension ceiling brings a sense of fluidity of the space.
2. Blooming lotus on the tiles and glass walls which touch the ground reflect each other.
3. Pedestrian path of black brick makes the people find themselves as if in real stone gate alley.
4. Small ornaments add color to the studio.

1	2	3

1．错层的设计弥补了平面空间带来的平淡感，
使空间变得更丰富更灵动。
2．凹凸有致的门楣别具风格。
3．镂空的吊顶与凹凸的砖墙搭配得相得益彰。

1. Alternate floor design offsets the plainness brought by plane space and makes the space clever and flexible.
2. The concavo-convex lintels are in special tastes and styles.
3. Engraved suspension ceiling and concavo-convex brick wall are in harmonious combination.

Wu Rui Decoration

五瑞装饰

位置：上海市金皖路389号金门广场106室　电话：＋86－021－50318310

随着众多现代派主义的出现，国内又兴起了一股复古风，那就是中式装饰风格的复兴。五瑞装饰的设计师追求的就是一种简练、现代的中式韵味，处处从细节入手，始终围绕房屋结构本身所特有的不同之处，使整个空间顿时活泼灵动起来。

五瑞装饰的大门设计非常独特，藏着一份犹抱琵琶半遮面的韵味。进入宽敞的大厅，明亮而又清爽。左边摆放着胡桃木色的家具，右边是颇具中式特色的屏风。薄薄的屏风，保持了空间良好的通风和透光率，营造出隔而不离的效果。屏风上描绘的名家书法更有种一气呵成的连贯和流畅。整个大厅赋予了空间深厚的内涵，沉稳中带着典雅。大厅的正前方就是五瑞装饰的工作室，吊顶设计和灯光装饰相统一，与整体风格遥相呼应，在功能上可用作照明使用，空间感更加突出。原木色的桌面上镶嵌着晶莹的玻璃马赛克，闪烁着个性的光芒。开放式的工作室突破了中式家具的呆板和严肃，同时是其他家具所难以比拟的。其传统中带有灵巧的线条，达到了观赏和使用的双重效果。

五瑞的会议室设计得也别具一格。半隔离的设计既保留了会议室的私密性，又强调了空间的通透性。盆景、书画及明清家具成为了整个空间的点睛之笔，营造出一种传统文化的氛围。西式玻璃茶几的简约并没有影响中式元素的发挥，两者相辅相成，配合得恰到好处，衬托出应有的气质和文化品位。

With the emergence of many modern styles, domestically a revival style populate is in vogue Chinese revivalism. The designers from Wu Rui are in pursuit of a sort of simple and modern Chinese charm.

The design of the gate of Wu Rui is very special, and it suggests a shy and eyesome taste. The hall is spacious, bright and cleanly with the walnut colored furniture on the left and the screen in Chinese style on the right. The thin screen contributes to sound aeration and light penetration, and an effect of partition yet inseparation is therefore created. Famous calligrapher's work on the screen is extremely fluent and coherent as if it is finished without any letup. The entire hall gives the space very profound meaning. The studio of Wu Rui is right in the front of the hall, and the suspension ceiling is in concord with lighting and echoes the entire style. Primitive wooden colored tabletop is furnished with translucent glass mosaic which radiates the rays of individualism. The open studio breaks through the stiffness and solemnity caused by Chinese furniture.

Wu Rui's meeting room is also very unique in style. Semi-isolation design keeps both the privacy and sense of being through. Bonsai, calligraphic and painting works, and furniture of Ming and Qing dynasties all become focal points of the space. These construct a kind of traditionally cultural atmosphere. The simplicity of glass end table in Western style doesn't affect the performance of Chinese elements. The two supplement each other and set off the deserved temperament and cultural taste.

P34-P35

	1
	2

1．设计独特的大门，藏着一份犹抱琵琶半遮面的韵味。
2．宽敞的大厅内摆放着胡桃木色的家具。

1. The design of the gate of Wu Rui is very special, and it suggests a shy and eyesome taste.
2. The hall is spacious, bright and cleanly with the walnut colored furniture on the left.

1	2	
	3	4

1．原木色的桌面上镶嵌着晶莹的玻璃马赛克，闪烁着个性的光芒。
2．开放式的工作室突破了中式家具的呆板和严肃，同时是其他家具所难以比拟的。
3．从屋内向外看，门窗宽大透光，采用天然材料，给人以宽敞明亮的清爽感觉，颇有和式风格。
4．薄薄的屏风，保持了空间良好的通风和透光率，营造出"隔而不离"的效果。

1. Primitive wooden colored tabletop is furnished with translucent glass mosaic which radiates the rays of individualism.
2. The open studio breaks through the stiffness and solemnity caused by Chinese furniture.
3. Large light-introducing doors and windows are made of natural materials, which give people a sense of brightness and coolness and are much of Japanese style.
4. The thin screen contributes to sound aeration and light penetration, and an effect of partition yet inseparateness is therefore created.

1	2	3

1. 吊顶设计和灯光装饰相统一，与整体风格遥相呼应。
2. 工作室的一边是一排排的书架。
3. 半隔离的设计保留了会议室的私密性，深色的沙发赋予空间深厚的内涵。

1. The suspension ceiling is in concord with lighting and echoes the entire style.
2. Rows of bookshelves are placed on a side of the studio.
3. Semi-isolation design keeps the privacy of the meeting room, and dark colored sofas give the space profound meanings.

1	2	3
		4

1．西式玻璃茶几的简约并没有影响中式元素的发挥，两者相辅相成，配合得恰倒好处，衬托出应有的气质和文化品位。
2．灰色砖墙的深沉和配饰的精美书画作品更使设计的文化内涵在不经意间得以体现，和谐自然。
3．墙上镶嵌的两幅画在柔和的灯光下渗透出浓浓的书香气息。
4．五瑞装饰一角。

1. The simplicity of glass end table in western style doesn't affect the performance of Chinese elements. The two supplement each other and set off the deserved temperament and cultural taste.
2. The profoundness of grayish brick wall and the refinement of the decorative handwritings and drawings embody the cultural meanings of the design harmoniously natural.
3. The two pieces of drawing effuse strong literary flavor.
4. A glimpse of Wu Rui.

Art Deco
凹凸库

位置：上海市莫干山路50号7号楼1楼　电话：+86－021－62778927

Art Deco（装饰艺术派），音译为凹凸库 ，建筑采用跃层设计、刚柔相济的横竖线条和丰富的装饰手法，并融汇了精致、内敛、闲适的气质，实现了凹凸库建筑艺术与都市生活的完美结合。

凹凸库采用了跃层设计，使原本平面单调的空间展现出自由灵动的氛围，同时通过窗子的不同比例，墙面的凹凸变化和楼高的跌落来强调建筑的变化和丰富；主体墙面采用仿木面通体砖墙，宽大的墙面饰以天然的木纹，整体建筑体现一种强烈的向上感；横竖线条简洁流畅，显现出建筑高贵而内敛。凹凸库的整个空间布局灵动，组合有机，步移景迁，处处营造着一种独特家居文化氛围，充分展现了一种动态均衡的构图，强调了几何图案之美。

宽敞的空间内，最多的装饰物就是沙发座椅，每一把都让人感觉异乎寻常的舒适。一楼的地面平铺着浅灰色的地毯，搭配富有传统色彩的家具，处处显示着典雅稳重。沿着木制材质配以合金扶手与护拦的楼梯向上，更让人不得不叹服凹凸库在细部装饰的细腻丰富，单是楼梯就体现了真正的刚柔相济。踩在二楼的地板上，感觉更多的是田园的闲适。家具虽与一楼并无太大的变化，但设计师却在二楼的地面采用了竖条纹的木板，深浅不一的木板配上木制的家具，田园气息迎面而来。

Art Deco architecture adopts alternate floor design, horizontal and vertical lines, plenty of decorative techniques, and some qualities such as refinement, introversiveness and leisure are melted into it. Adopting alternate floor design makes the space free and lively, and meanwhile through different proportions of the windows, concavo-convex surface of the walls and different heights of buildings, diversification and richness of architecture are created. The walls of the main body use wooden-mimic bricks, the effect of the whole piece of wall is like natural grains of wood, and the entire architecture conveys a strong sense of rising-up. The horizontal and vertical lines are simple and fluent, which infuses the architecture with nobility and introversiveness. The entire space of Art Deco is lively and smartly laid out with harmonious combinations and views against positions. Special domestic cultural atmosphere can be sensed everywhere and dynamic equivalence and geometrical beauties are emphasized.

The most ornaments are sofas within the capacious space and each of them makes people feel extremely comfortable. Light grayish carpet is furnished on the floor of the first floor. The furniture in traditional colors are very elegant and steady. The wooden balustrade and alloy handrails of the staircase reveal the richness of the decorative detail. On the flooring of the second floor, what people feel is villatic quietness. The furniture here is almost the same with the one on the first floor, but vertical grain pattern is adopted on the flooring of the second floor. Villatic taste is obvious within the space where here is the combination of the wooden floor and wooden furniture.

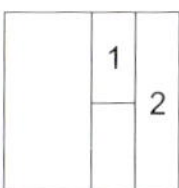

1．木制材质的楼梯配以合金扶手与护拦，体现了真正的刚柔相济。
2．跃层的设计，使原本平面单调的空间展现出自由灵动的氛围。

1. Wooden balustrade and alloy handrails of the staircase bring out a combination of masculinity and femineity.
2. Alternate floor design turns the formerly monotonous space into a free and flexible one.

P44-P45

1．二楼的地面采用了竖条纹深浅不一的木板。
2．一楼的地面平铺着浅灰色的地毯，搭配富有传统色彩的家具，处处显示着典雅稳重。
3．深浅不一的木板配上木制的家具，田园气息迎面而来。
4．主体墙面采用仿木面通体砖墙，宽大的墙面饰以天然的木纹，整体建筑体现一种强烈的向上感。

1. On the second floor, the flooring is in vertical grains of dark and light colors.
2. Light grayish carpet is furnished on the floor of the first floor and the furniture in traditional colors is very elegant and steady.
3. Villatic taste is obvious within the space where there is the combination of the wooden floor and furniture.
4. The walls of the main body use wooden-mimic bricks, the effect of the whole piece of wall is like natural grains of wood, and the entire architecture conveys a strong sense of rising-up.

BICHE de BERE
碧雪德贝尔

位置：上海市南苏州路1295号 电话：+86－021－63580177

碧雪德贝尔的设计风格是古朴与现代的完美结合。设计师使用独特的错层设计将整个空间分割为两部分，休闲区以及办公陈列区。这样的设计打破了室内空间平层的呆板乏味，使得动静分区更加明显，让各个功能区间错落有致，极富层次感。由于错层而被抬高的休闲区，零星散放着风格独特的藤制休闲椅及西式休闲椅，仅三层的台阶就增加了此区域的私密性，也增强了采光功能。

办公陈列区主要由办公区及陈列区组成。办公区强调了现代性与舒适性，更符合办公需求。无论是橙色的贵妃沙发、玻璃桌面，还是铝合金座椅，皆在体现现代的时尚感，搭配长条形的仿古色木制地板，古朴中带着现代的美感。尤为值得一提的是设计师将木格子与工艺玻璃、装饰布等材料联合运用，形成了一个半隔离的玻璃房，集传统与现代为一体，通透中又不失装饰性。与办公区相比较，陈列区就显得朴素自然。整个区以红色的块毯作为分割，色调一致，风格统一。藤制的座椅色彩幽雅，风格清新质朴，为整个空间营造出一种朴素的自然气息。

碧雪德贝尔的点睛之笔更在于设计师大胆地运用横竖交叉的直线组合成的房梁，人为制造出一种典雅及通过排列组合后的韵律及工艺美感。既拉伸了空间，也使整个工作室更具立体感。

The design style is a perfect combination of primitive simplicity and modernism.Through unique alternate floor design the designer divides the entire space into two: leisure area and office area. Such design breaks the stiffness and monotony of the interior space, smartly classifies different functions of various parts of the space. Some rattan leisure and western styled chairs are randomly placed in the raised leisure area due to alternate floor design. Three step tiers contribute privacy and lighting function to this area.

Office area consists of office section and displaying section. Office section emphasizes modernism and comfort, which accords with the need of official business. The orange sleeping sofa, glass tabletop and alloy made chair all show modern vogue. Archaized grained wooden flooring is both primitively simple and beautiful. The worth of mentioning is that the designer put wooden lattice, artistic glass and decorative textiles together to form a semi-isolative glass chamber which is a mix of tradition and modernism. In contrast, the displaying section seems austere and natural.

Smartest point of BICHE de BERE lies on the bold use of intersectional horizontal and vertical beams, which deliberately creates a sort of elegance, rhythm and artistic beauty through permutation and combination. The space is lengthened and the entire studio is more dimensional.

P46-P47

	1
	2

1．独特的错层设计，打破了室内空间平层的呆板乏味，使得动静分区更加明显，让各个功能区间错落有致，极富层次感。
2．被抬高的休闲区，增加了私密性，也增强了采光功能。

1. Unique alternate floor design breaks the stiffness and monotony of the interior space, smartly classifies different functions of various parts of the space.
2. The raised leisure area due to alternate floor design contributes privacy and lighting function to this area.

1	2

1．设计师将木格子与工艺玻璃、装饰布等材料联合运用，集传统与现代为一体，通透中又不失装饰性。
2．设计师大胆地运用横竖交叉的直线组合成的“格子”，人为制造出一种典雅及通过排列组合后的韵律及美感。

1. The designer put wooden lattice, artistic glass and decorative textiles together to form a semi-isolative glass chamber which is a mix of tradition and modernism.
2. The designer boldly uses intersectional horizontal and vertical lines, deliberately creates a sort of elegance, rhythm and artistic beauty through permutation and combination.

P50-P51

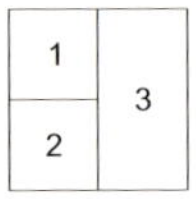

1. 长条形的仿古色木制地板搭配橙色的西式休闲沙发，古朴中带着现代的美感。
2. 纵横交错的房梁拉伸了空间，也使整个工作室更具立体感。
3. 玻璃的灯罩、玻璃的盛具、玻璃的桌面，为厚重的木制地板与房梁增加了通透感。

1. In combination with Archaized wooden grained flooring is both primitively simple and beautiful.
2. Beams in length and breadth extend the space and make the whole studio three dimensional.
3. Glass lampshade, wares and desktop add sense of penetration to the wooden flooring and beams.

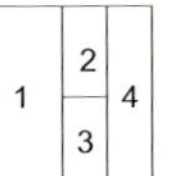

1. 藤制品色彩幽雅，风格清新质朴，为整个空间营造出一种朴素的自然气息。
2. 陈列区以红色的块毯作为分割，色调一致，风格统一。
3. 横竖交错的陈列架与房梁遥相互应。
4. 风格简约的展台。

1. Rattan chairs are graceful in color and fresh in style, and form an inornate natural atmosphere for the space.
2. Red rugs play a role as dividing space in display area.
3. Display shelves are in concord with the beams.
4. Simply styled cultural wall.

CA-GROUP Bookshop
文筑书店

位置：上海市南苏州路1305号1楼　电话：+86－021－63273381

文筑书店中最突出的装饰是书，最简单的装饰也是书。可以说，书就是这里的灵魂。在这里，书不但是装饰品，更是工作室设计布局的灵感来源和装饰元素。而放置书的书柜、书橱和书架，就成了工作室内空间分隔、工作室布置和功能分区的最直接的设计元素。

步入文筑书店，入口处浅灰色的外墙斑驳地裸露着白色的砖体，粗犷的墙体显得很是热情。迎面便是一个顶天立地的开放式书架，上面整齐地摆放着多本人文及传统文化类图书，使人对工作室的定位及特色先入为主，有了感性认识。入口大厅是二层楼高的挑空，中间没有任何阻隔，单单留着几根锈迹斑斑的支架，使空间的立体感更加丰富，整个建筑线条现代简洁、个性流畅。左边落地外框式玻璃幕墙采光通透，使整个工作室感觉更宽敞明亮。右边的工作室继续了开放式书架的设计风格。除了中间的书桌，周围几乎全是书橱，视觉上，顶天立地的书橱代替了墙壁的所有装饰，工作室感觉更像一个空间紧密、布局合理的图书馆阅览室，整个室内充满文化气息。

文筑书店就工作室的功能而言，做得相当简洁。将书橱作为墙体，一方面合理利用空间，另一方面也方便了工作者的取阅。中间长条形的书桌及旁边的小圆桌，都为工作者提供了良好的工作场所，处处体现出人性化的设计。

The most outstanding ornaments are books, and so are the simplest ones. Books are the soul here. Books are not only decoration but also the source of designing inspiration and elements. The bookcases, book cabinets and bookshelves become the partition elements of the interior space and of the functions.

At the entrance of CA-Group Bookshop, light grayish wall reveals white bricks, which makes the rough wall block passionate. In straight front is a piece of ceiling reaching bookshelf, on which many books about humanities and traditional cultures are placed. People can directly and perceive the purpose and feature of this very studio. The hall at entrance is a raising space two floors high without any blocks but some brackets full of rusting stains, which makes the entire studio dimensional, the entire architectural lines simple, modern and characteristic. The grounded external framed glass curtain on the left is well day-lighted and makes the entire studio more spacious and bright. The studio room on the right succeeds to the style of opening bookshelf. Except for the desk in the middle, the book cabinets are almost around the room. Visually, the ceiling reaching bookshelves take the place of all wall decorations. The studio is felt more like a space compact, reasonably laid out reading room of a library where there is full of cultural taste.

As to its function, CA-Group Bookshop is very simplified. Bookshelves as wall blocks is both for the sake of good utilization of the space and for the sake of readers' convenience of reading. The strip shape desk and small round desks nearby all provide good places for the people and the human designs are evident everywhere.

EL croquis

P54-P55

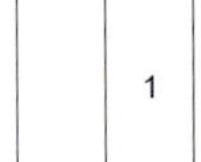

1．浅灰色的外墙斑驳地裸露着白色的砖体，入口处“顶天立地”的开放式书架线条现代简洁、个性流畅。

1. The light grayish wall reveals white bricks, and ceiling reaching bookshelves at the entrance are concise in lines and are characteristic.

1．“通天”的书柜，全部做成了30 cm见方的格子，取用非常方便。
2．顶天立地的书橱代替了墙壁的所有装饰，工作室感觉更像一个空间紧密、布局合理的图书馆阅览室，整个室内充满文化气息。

1. The ceiling reaching bookcases are all composed of 30x 30cm squares, which are convenient for fetching books.
2. The ceiling reaching bookcases substitute for all wall decorations. The studio is felt more like a space compact, reasonably laid out reading room of a library where there is full of cultural atmosphere.

The Art House
艺术坊

位置：上海市北苏州路1305号底层　电话：+86－021－63276430

艺术坊的设计运用了大量古代建筑的元素，力求体现工作室的反璞归真。整个空间布局合理，利用充分，将工作室所应具备的功能与特性全部融入到了设计中。

入口处，简洁隽秀的的古代装饰桌静静地靠墙而立，两边对称摆放着瓷器花瓶，浓浓的古朴与沧桑气息透过这些装饰物扑面而来。飘逸的"THE ART HOUSE 艺术坊"几个大字醒目地勾勒于墙面上。入口的右边"顶天立地"地矗立着一块浅灰色的石碑，上面清晰雕刻着孔子的一句佳句："有朋自远方来，不亦乐乎。"古代家具配以古人诗词，古朴自然，又不失精巧典雅。

再往里走，艺术坊的工作室就尽收眼底。整个工作室设计得非常巧妙，虽然空间有限，设计还受到了房柱的影响，但设计师却合理利用了每一块空间，让瑕疵也变成了设计的亮点。这里每一间工作室都是半隔离的设计，每间工作室之间的间距就是房柱与房柱之间的距离。这样的设计既避免了房柱的尴尬，又保证了每一间工作室的独立性。同时，保留了木制材料所原有的木纹及瑕疵的原木色房梁与房柱，使得设计风格更加自然随意。

工作室的房顶设计也独具匠心，线条形的设计层层叠进，在视觉上，拉伸了整个空间，使整个空间陡然之间宽了许多，设计之巧妙真是不言而喻。

The design of the Arts Workshop uses plenty of ancient architectural elements, and it does its best to achieve the effect of being primitively simplified and real. The entire space is reasonably laid out and sufficiently utilized in that all functions and characteristics a studio deserves are all included in the design.

At the entrance, simple and graceful antique table stands quietly against the wall, and symmetrically, two ceramic vases are respectively placed on each side of the table. Strong senses of primitive simplicity and change in world fall upon people. The elegant "THE ART HOUSE" is remarkably engraved on the wall. At the right side of the entrance stands tall and upright a piece of monolithic stone monument in light gray on which engraved a motto from Confucius: "Friends come from afar, happiness falls upon heart". Antique furniture in companion with historical figure's verse seems not only primitively and natural, but also refined and elegant.

As going inside further, the studio of the ART HOUSE is in full scope of one's eyes. The entire studio is very artful in design. Although the space is limited and the design is restricted by the columns within the house, the designer reasonably utilizes every part of the space and converts flaw into highlight. Every studio is semi-partitioned, and the distance between each two studios is identical to that between two columns. The design of this kind both avoids the defect of column's presence and assures the independence of each studio. Meanwhile, wood grains and flaws of the columns and beams are reserved and this makes the design more natural and casual.

The ceiling of the studio is masterful and crafty too, and line patterned design visually lengthens and broadens the entire space.

1. 简洁隽秀的古代装饰桌静静地靠墙而立，两边对称摆放着瓷器花瓶，浓浓的古朴与沧桑气息透过这些装饰物扑面而来。
2. 一块浅灰色石碑上雕刻着孔子的一句佳句：“有朋自远方来，不亦乐乎。”

1. Simple and graceful antique table stands quietly against the wall, and symmetrically, two ceramic vases are respectively placed on each side of the table. Strong senses of primitive simplicity and change in world fall upon people.
2. A piece of monolithic stone monument in light gray on which engraved a motto by Confucius: “Friends come from afar, happiness falls upon heart.”

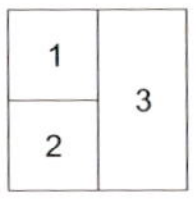

1	3
2	

1．原木色的房梁与房柱保留了木制材料所原有的木纹及瑕疵，显得更加自然随意。
2．线条形的房顶设计层层叠进，在视觉上，拉伸了整个空间。
3．隔而不离的工作室设计巧妙。

1. Wood grains and flaws of the columns and beams are reserved and this makes the design more natural and casual.
2. Line patterned ceiling design visually lengthens and broadens the entire space.
3. Studio in smart design: partition against inseparateness.

NewsDays
集美组

位置：杭州市里鸡笼山88号 电话：+86－0571－87975060

集美组是一个典型的底层复式工作室，设计师在设计时不但满足了工作室对于功能空间的要求，而且解放出了很多令人感觉尴尬的空间，使空间显得更加顺畅、开阔，挖掘出了复式结构在空间方面天生优越的本能。为了使整个空间看上去更加宽敞明亮，设计师使用最简单的白色作为设计的主色调。白色的墙、白色的灯饰、白色的屋顶，整个空间一气呵成，使工作室显得更加整洁宽敞。同时，设计师还以“阳光”作为设计的主题，力求创造一个可以给予灵感的工作空间。白色的主调，配以原木色的家居，金色的阳光透过高挑的落地玻璃窗及斜顶的玻璃窗洒满了整个室内空间，这样的工作环境让人舒心又静心。

集美组总体的设计风格基本围绕简约二字。一楼大厅的一张造型独特的书桌让人眼前一亮，这是由合金、玻璃与木材搭配而成的，凌乱的桌脚非但没有让人觉得眼花缭乱，反而透着一股简单美，颇具有现代艺术的气势。拾阶而上，从二楼平台俯视，立刻就能感受到正厅那澄净明亮的气息。二楼的入口是一个半圆拱形砖砌拱门，在厚重的白色文化墙的衬托下，颇有点古堡式建筑的风格。设计师在二楼的过道处作了一番巧心布局，陡斜的屋顶处设计了一扇玻璃窗，阳光一泻千里。同时利用半圆形的拱门将整个空间连接在一起，呈现出了彼此独立又相互流通的节奏感。

NEWSDAYS is traditionally a lowly storied complex studio, the design of which not only satisfies the requirement of functional spaces from a studio, but also frees some spaces which make people embarrassed. Therefore, the spaces seem more fluent and spacious, and the superiority of complex space is developed. In order to make the entire space more spacious and bright, the designer uses the simplest white as the major color. The white wall, white lighting accessories and white ceiling make the entire space coherent and the studio more cleanly and spacious. Meanwhile the designer takes “sunshine” as the theme of the design in order to create a working space which can inspire people within. A space with the white color as major hue, originally wood-grained furniture and golden sunshine pouring down makes the working environment both comfortable and tranquil.

The general style of NEWSDAYS is about simplicity. What attracts people's eyes at the hall of the first floor is a uniquely shaped desk which is made of alloy, glass and wood. The desk legs are simple and full of modern artistic taste. Taking a bird's eye view from the second floor, one can immediately feels the exclusive brightness. The entrance at the second floor is a semicircle bricked arch which is set off by white cultural wall, and it is full of castle style. At the aisle of the second floor, the designer makes an artful design, of a window on the steep roof, through which sunshine can travel smoothly in. With the arch, the spaces are jointed and a sense of both independence and communication occurs.

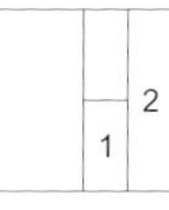

1．合金、玻璃与木材的合理搭配，具有现代艺术的气势。
2．复式的结构，整个空间宽敞明亮。

1. The combination of alloy, glass and wood bears modern artistic taste.
2. Complex space is spacious and bright.

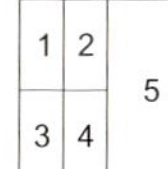

1、2. 过道开阔明亮，墙面布置简单大方。
3、4. 拾阶而上，从二楼平台俯视，立刻能感受到正厅那澄净明亮的气息。
5. 厚重的白色文化墙、半圆拱形砖砌拱门、陡斜的屋顶，颇具古堡式建筑的风格。

1、2. The aisle is bright and wide; the wall is simply furnished.
3、4. Viewing from the second floor, one can sense clearness and brightness of the hall.
5. Thick white cultural wall, semicircle bricked arch and pitched roof contribute to somewhat castle style.

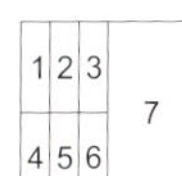

1. 高挑的落地玻璃窗让工作室与阳光亲密接触。
2、5. 半圆的拱门设计是这个工作室的设计亮点。
3. 陡斜的屋顶。
4. 考究大气的布局，令空间充满着律动的节奏感。
6. 色调以原木色和白色为主，整个空间一气呵成。
7. 白色的墙、白色的灯饰、白色的屋顶，使整个工作室显得整洁宽敞。

1. Tall grounded glass windows allow intimate contact between the studio and sunshine.
2、5. Semi-circle arch is the highlight of the studio design.
3. Pitched ceiling.
4. Generous layout makes the space full of rhythms.
6. Wood grain color and white are dominant colors, and the space is coherent.
7. White walls, light ornaments and ceiling make the whole studio clean and spacious.

LWM Architects+Shanghai Yue Jie

李玮珉建筑事务所+上海越界

位置：上海市七莘路3758号 电话：+86－21－51085585

李玮珉建筑事务所是一个二层楼的办公楼，整体设计采用全框架可自由组合分割的结构。遵循现代办公的要求，利用了适当比例的空间规划与色彩搭配，跳脱出传统办公室拥挤杂乱、制式刻板的形态。加上简洁大方的整体线条、协调的颜色搭配，以及机能完善的现代设备，凸显气派、尊贵和舒适。

走进一楼大厅，是一个宽敞的接待区，右边的会议室采用了整幅的落地玻璃墙，强调了体积感和通透性。左边是办公区，李玮珉建筑事务所设计的最大亮点在于它巧妙地利用了这幢办公楼的高度；设计师将作为办公区的一楼和二楼打通，创造了一个超高的办公空间。这样的设计既开阔了视野，又显得气势恢弘，显示出整幢办公楼的气派。开放式的办公区按照现代办公模式被分割成模块式的办公空间，充分满足了现代办公的人性化要求。层次分明的吊顶设计，不仅使原本单调的空间更富有层次变化，还提升了空间质量。整个一楼的设计强调办公的开敞与组合，主管区、接待区、会议区围绕着大办公空间，通过办公家具的空间设计使环境显得秩序井然。

李玮珉建筑事务所的办公室在入口处及最靠左边的财务室都设计了通往二楼的楼梯，在这个楼梯中，设计师摒弃所有的装饰，使楼梯腾空在大厅之间，折叠向上的曲线充分体现了现代时尚感，别致又实在。二楼的整体设计延续了一楼的风格，简洁明快，高效时尚。设计师在办公区的一侧设计了图书室，充分满足了作为工作室的功能需求。

LWM Architects is a two storied office building, and its design uses full framework which is freely detachable and combinative structure. According to modern office requirement, proper proportioned space plan and color group are used here and the conditions of being jammed, chaotic and rigid in form are avoided.

There is a spacious reception area in the hall of first floor. The meeting room on the right uses grounded glass walls which emphasizes the sense of volume and sense of being through. The highlight of the design of LWM Architects lies on the artful utilizing the height of this office building. The block between first and second floors is get through, therefore a super high office space is created. Such design both broadens the vision and makes the entire building more magnificent. The opening office area is divided into module parts according to modern office pattern and satisfies the modern office's human need. The smart ceiling design not only makes the usually monotonous space more diverse, but also improves the quality of the space. The design of the first floor emphasizes the opening and combination of offices. The administration area, reception area and meeting area surround the office space, where the arrangement of much office furniture makes the environment orderly.

Staircases leading to the second floor are set at the entrance of the office and near the financial section room on the far left. The designer disposes of all decorations and lets the staircases running between the halls, zigzag upward; the curvatures of the staircases fully show their modern genre. The design at second floor succeeds to that of the first floor: simple, bright, effective and voguish.

1/306 外观立面图

2/306 外观立面图

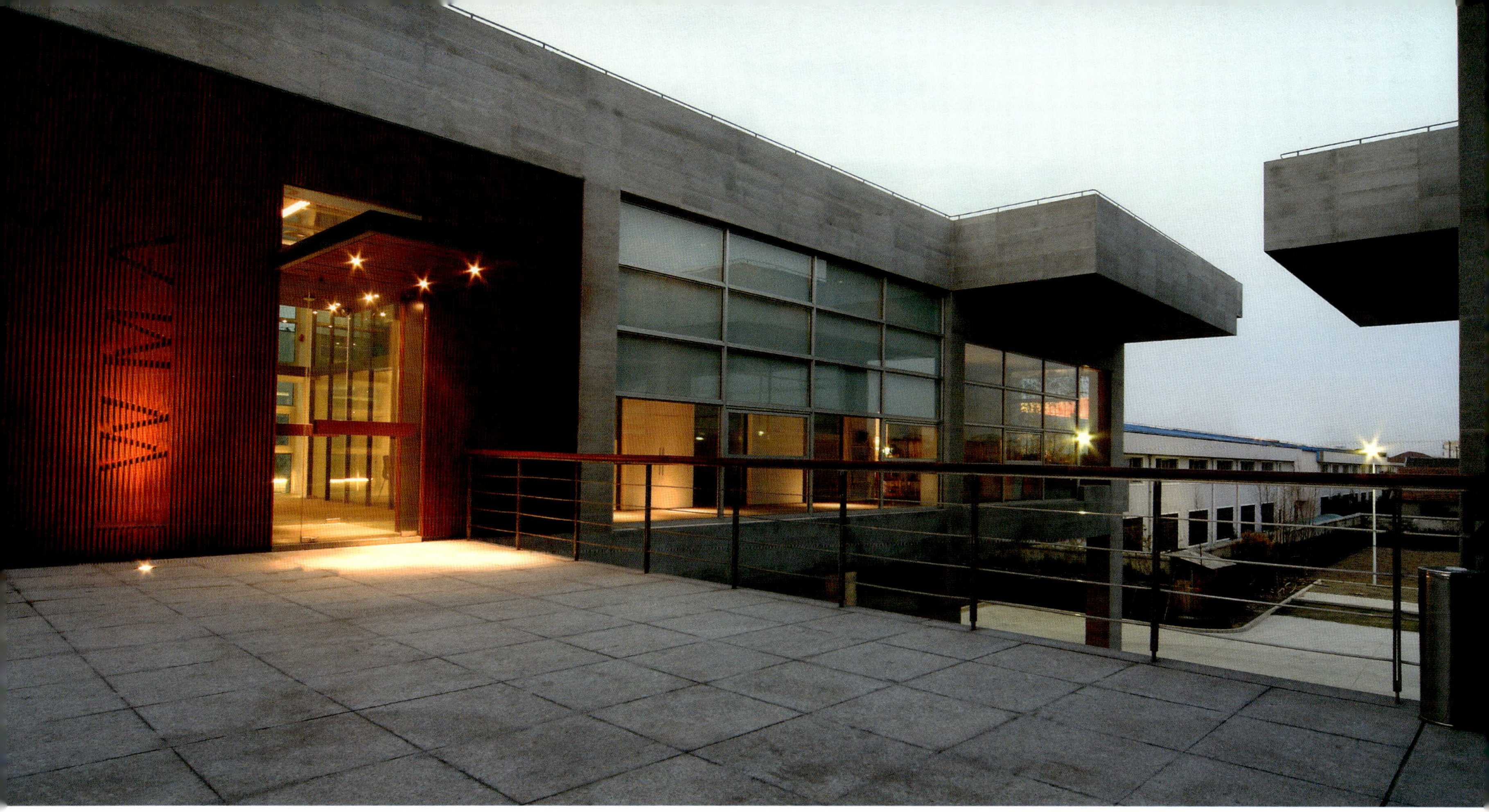

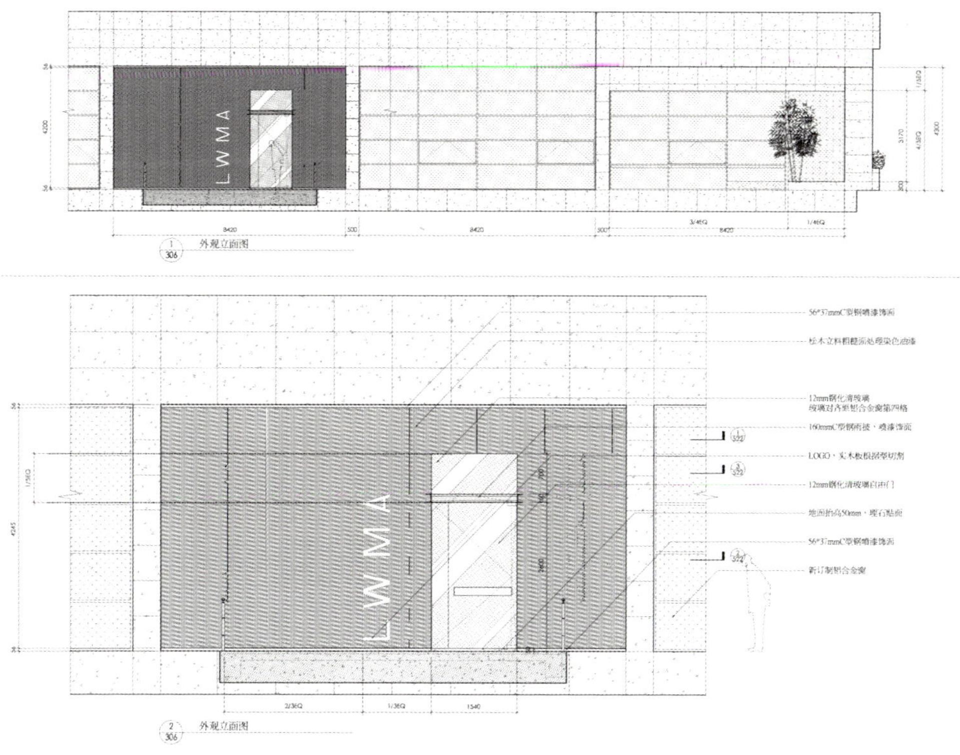

1

1．从二楼远望上海惠宝办公室，更显气派尊贵。

1. Being viewed from on the second storey, Shanghai Huibao Office seems more noble and luxurious.

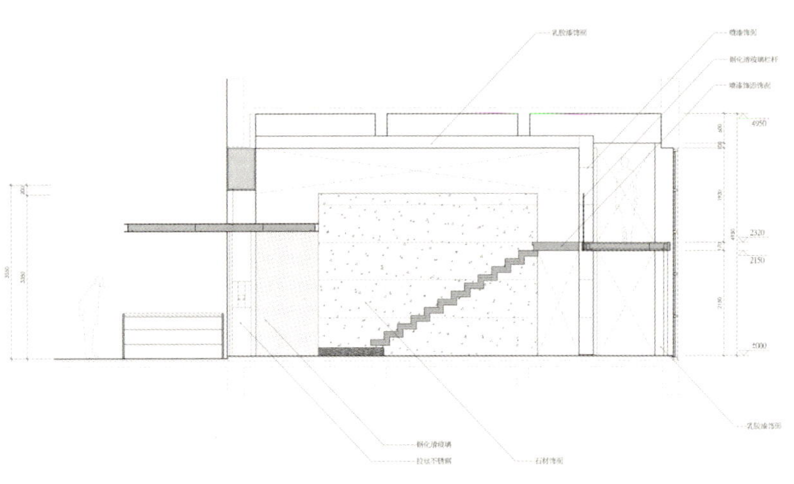

乳胶漆饰面
喷漆饰面
钢化清玻璃栏杆
4950
2320
2150
±000
乳胶漆饰面
钢化清玻璃
拉丝不锈钢
石材饰面

1	3
2	

1．在这个楼梯中，设计师摒弃所有的装饰，
　使楼梯腾空在大厅之间。
2．整幅的落地玻璃墙气势恢弘。
3．线条简朴的玻璃门，体积感强，透视
　效果好。

1. The designer disposes all decorations and lets the staircases running between the halls.
2. The whole piece of grounded glass wall is sublime.
3. Glass door in simple lines is strong in cubicity and in perspective.

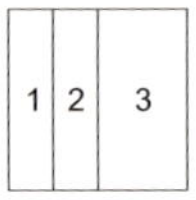

1．层次分明的吊顶设计，不仅使原本单调的空间更富有层次变化，还提升了空间质量。
2．折叠向上的曲线充分体现了现代时尚感，别致又实在。
3．办公区设计成可自行分割的大开间办公室及会议室，满足了办公及会议的要求。

1. The smart suspension ceiling design not only makes the usually monotonous space more diverse, but also improves the quality of the space.
2. Zigzag upward; the curves of the staircases fully show their modern genre.
3. The office area is designed as large office room or meeting room that can be freely divided, which meets the requirement of office work and meeting.

1	2

1．一层的玻璃办公区。
2．在柔和的灯光下，大幅玻璃墙没有平淡感。

1. Office area enclosed in glasses on first storey.
2. The glass walls in large pieces are not plain under soft lighting.

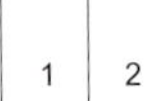

1．层次分明的吊顶设计，使原本单调的空间更富有层次变化，还提升了空间质量。
2．设计师将作为办公区的一楼和二楼打通，创造了一个超高的办公空间。

1. The smart suspension ceiling design not only makes the usually monotonous space more diverse, but also improves the quality of the space.
2. The office area between first and second floors is get through, therefore a super high office space is created.

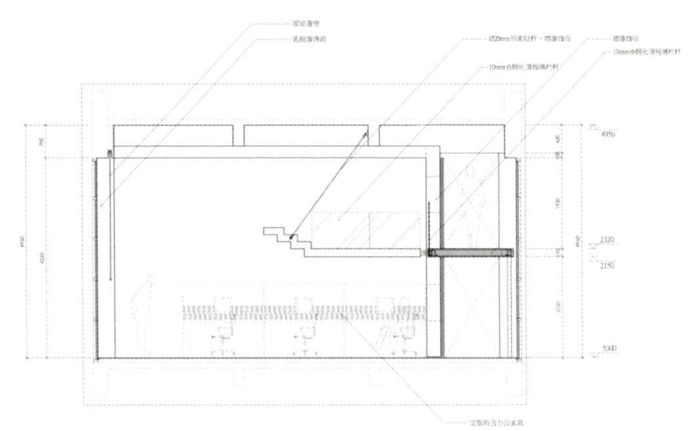

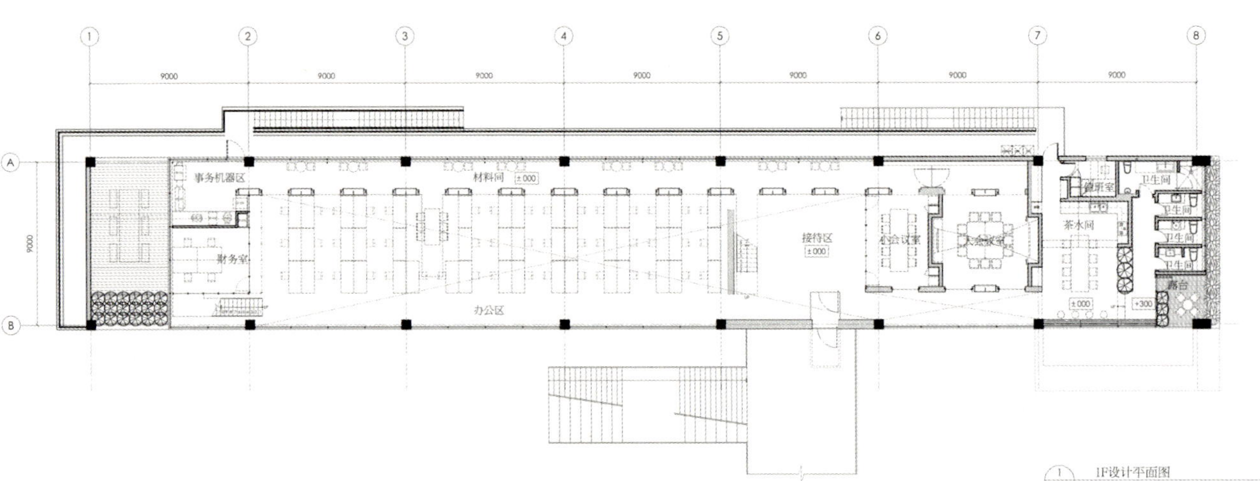

1/201 1F设计平面图

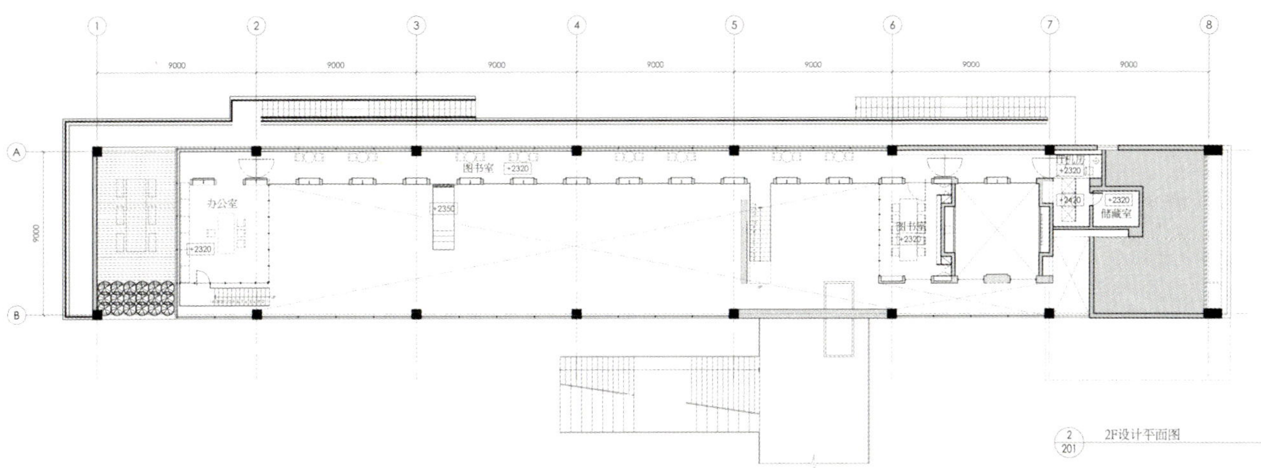

1
2
3
4
5
6
7
8
9000
9000
9000
9000
9000
9000
9000
A
B
图书室
+2320
办公室
+2320
+2350
+2320
+2420
+2320
储藏室
2
201
2F设计平面图

P78-P79

1	2	3	4

1. 二楼的图书区。
2. 落地玻璃窗增添了办公区的通透性。
3. 开放式的办公区按照现代办公模式被分割成模块式的办公空间，充分满足了现代办公的人性化要求。
4. 整体设计的风格，简洁明快，高效时尚。

1. Library area on second storey.
2. grounded glass window increase the sense of penetration for the office area.
3. The opening office area is divided into module parts according to modern office pattern and satisfies the modern office's human needs.
4. The entire design style is concise, bright, effective and fashionable.

1	2	3	4

1、2. 别具一格的挑空设计。
3、4. 高技术的会议设施，装饰简洁、得体。

1、2. Unique styled high-rising design.
3、4. Facilities centered on high technique are simply furnished.

Loewe's Trade
立文贸易

位置：南通市国贸大厦8楼　电话：+86－0513－3580100

立文贸易度采用的是一种e时代的办公模式，强调多样性和个性化。它选用新材料、新工艺、新设施，把办公楼装修得更像是一个休闲场所。让人置身其中，真正感受“色彩有致，空间有韵”。

在本案的设计中，设计师大量采用有机的形态作为设计元素，成功地柔化了呆板的建筑空间。在色彩与图案的表现上运用了多样性的手段，采取优秀的配色与图案计划，将色彩的调和及必要的对比加以充分的考虑，使不同区域的色彩吻合工作环境的氛围。同时，在色彩及用光上强调了冰激凌色与玻璃及金属的搭配，使环境呈现出一种强烈的对比与虚幻的境界。

在功能划分上，整个办公楼被分隔成会议室、就餐区、休闲区以及独立的工作室。每个区域的设计都有其独特的地方。独立工作室的设计相当的现代，椭圆的金属桌面、黑色的休闲椅、白色的书橱、透明的玻璃搁架，功能齐全，简约却不简单。在就餐区的设计中，将长方形的空间加以曲线的柔化处理，并饰以具有快餐特色的鲜明色彩，从而形成一个活跃、时尚的就餐环境。设计师大胆地启用冰激凌色的透明隔离墙，五彩斑斓的色彩活跃了工作室，凸显个性化的时尚。同时，设计师更以玻璃及灯光打造出一个虚幻的场景，不仅在视觉上开阔了整个办公楼，更让人时刻体验色彩的光怪陆离。

Loewe`s trade uses a kind of e-time office mode, which emphasizes diversity and individuality. It uses new material, new techniques, new equipment and facilities and furnishes the office building in a way that makes it more like a leisure place where people truly sense the fine colors and rhyme of space.

In the design, the designer plentifully uses organic forms as design elements and successfully softens formalistic architectural space. In expression of colors and patterns, the designer uses diverse means and uses excellent match colors and pattern design. The concord and contrast of colors are put into full consideration and colors in different areas match the different working environment. Meanwhile, in colors and lightings, ice cream color is in companion with glass and metal, which makes the space an environment of strong contrast and illusion.

In terms of functions, the entire building is divided into meeting room, dining area, leisure area and independent studio rooms. Each area has its specialty. The studio rooms are very modern in design, elliptical metal desktop, black leisure chairs, white book cabinets and transparent glass shelves are all ready in function, simplified yet not simple at all. In dinning area, the oblong space is softened by using curvatures, bright colors apt to dinning are used to contribute to an active voguish dinning environment. The designer boldly uses ice cream colored transparent partition wall, through which various colors from without enrich the studio and individual vogue is shown off. Meanwhile, the designer uses glass and lighting to create an illusive scene which not only broadens the vision of the entire office building space, but also enables people to sense the gaudiness.

1. 设计师以玻璃及灯光打造出一个虚幻的场景，不仅在视觉上开阔了整个办公楼，更让人时刻体验色彩的光怪陆离。

1. The designer uses glass and lighting to create an illusive scene which not only broadens the vision of the entire office building space, but also enables people to sense the gaudiness.

1	2	3

1．椭圆的金属桌面、黑色的休闲椅、白色的书橱、透明的玻璃搁架，功能齐全，简约却不简单。
2．独立工作室的设计相当的现代。
3．设计师大胆地启用冰激凌色的透明隔离墙，五彩斑斓的色彩活跃了工作室，凸显个性化的时尚。

1. Elliptical metal desktop, black leisure chairs, white book cabinets and transparent glass shelves are all ready in function, simplified yet not simple at all.
2. The studio rooms are very modern in design.
3. The designer boldly uses ice cream colored transparent partition wall, through which various colors from without enrich the studio and individual vogue is shown.

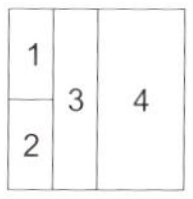

1．工作室的色调以白色为主，最大化地满足了工作室的使用功能，又不失干净、利落、简洁、自然的个性。
2．宽畅的工作空间，白色的墙，白色的百叶帘，再配以深色系的办公椅，色彩分明，干净利索。
3．设计师在色彩与图案的表现上运用了多样性的手段。
4．在色彩及用光上强调了冰激凌色与玻璃及金属的搭配，使环境呈现出一种强烈的对比与虚幻的境界。

1. White is the dominant color of the studio, and it maximumly satisfies the functions of the studio and takes with cleanly, brisk, simple and natural character.
2. Loose working space with white walls and blinds and dark colored office chairs are contrasty in colors and cleanly.
3. The designer uses diverse means and uses excellent match colors and pattern design.
4. In colors and lightings, ice cream color is in companion with glass and metal, which makes the space an environment of strong contrast and illusion.

1 2 3

1．强调曲线的意义也是改善办公环境的一种手法。由于曲线的柔和及运动的效果，将或多或少地减少工作的单调乏味。
2．因地制宜制作的彩色玻璃墙，恰到好处地填补了白色带来的单调感。
3．将长方形的空间加以曲线的柔化处理，并饰以具有快餐特色的鲜明色彩，从而形成一个活跃、时尚的就餐环境。

1. Emphasis on curve is one of the methods to improve the environment. Owing to its softness and dynamic effect, the monotony of works will be more or less reduced.
2. Colorful glass walls offset the monotony brought by white.
3. The oblong space is softened by using curves; bright colors apt to dinning are used to contribute to an active voguish dinning environment.

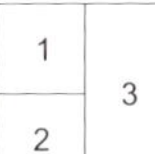

1．时尚的休闲区一角。
2．五彩缤纷的透明隔离墙为整个工作室带来一丝灵动。
3．体验色彩所带来的魅力。

1. A glimpse of fashionable leisure area.
2. Colorful transparent partition walls bring a sense of cleverness to the studio.
3. Charm brought by colors.

P92-P93

1．白色的装饰墙。
2．运用色彩与光营造科技的场所感。
3．柱与背景的光亮材料是典型的光亮派手法。
4．造型奇特的会议桌。

1. White decorative wall.
2. Exertion of colors and light contribute to a place of high science and technology.
3. Columns and background lights are typical manner of brightness pursuers.
4. Fantastic formed meeting board.

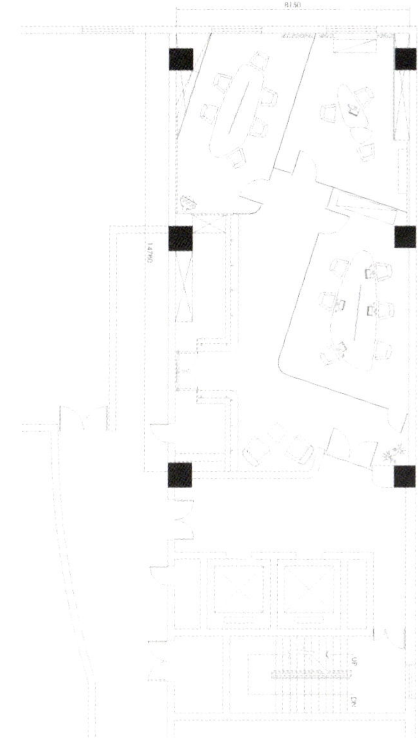

Interior Architecture Design

内建筑

位置：杭州市拱墅区杭印路49号6号楼 电话：+86－0571－88238832

内建筑的设计师力求创造的是一个艺术而有深度感的空间。整个设计风格是一个典型的创意仓库，老仓库的历史底蕴、高大空阔的房屋结构，为设计师的设计提供了非常大的灵活性。在这个设计中，设计师保留了许多老建筑的个性元素：斑驳的砖墙、灰暗的水泥地面以及裸露的管道。原始的风格搭配现代西式的办公家具，厚重中带着简洁，粗犷中带着细腻。

设计师利用了仓库特有的高大灵活，打破了原来的结构，用厚重的钢铁重组了这个自由空间，将整个工作室设计成了跃层，模糊了传统的空间功能区分，使整幢房子的室内变得更灵活。同时根据不同的需要，把工作室分割成若干个相对独立的空间。由于保留了旧仓库的元素，整个工作室给人的感觉是极端张扬的。在这个宽敞的工作室内到处可见裸露的管道和阀门，纵横交错地充塞了整个视线。斑驳的砖墙，粗犷而又热情。黑漆漆的带着红色锈斑的钢铁高唱着主旋律。线条简洁流畅、五颜六色的西式办公家具更是大张旗鼓。

设计的亮点更在于每个独立空间的独特风格。黑漆漆的钢板上，火红色的西式沙发与空中悬挂的厚重钢铁管道搭配得相得益彰。透明落地玻璃窗制造的悬空工作室更让人拍案叫绝。设计师巧妙地将这些冲撞的元素搭配在一起，使整个空间蕴涵个性化的审美情趣，更为工作者创造了绝好的创作空间。

The designer of Interior Architecture Design tries his best to create an artistically deep space. The entire design style is traditionally originative storehouse whose historical containing and tall and broad structure give the designer much flexibility. In this design, the designer reserves many individualistic elements of old architectures: stained brick walls, dark cement ground and exposed pipelines. Its original style is in match with modern western-styled office furniture: massiness versus simplicity, and roughness versus subtlety.

The designer utilizes the tallness and flexibility of the storehouse and breaks the original structure to use bulky steel to reform this free space, to design the studio into alternate floor pattern and to blur the traditional functional space section, which makes the entire house more flexible. Meanwhile, according to different needs, the studio is divided into several independent parts. Because the elements of the old storehouse are reserved, the entire studio gives people a sense of extension. Exposed pipelines and valves can be seen everywhere within this spacious studio and stained brick walls are rough and passionate. Dark steel with rusting stains play dominant role in people's visions, and colorful simple-lined western-styled office furniture are showing themselves off.

The highlight of the design lies on unique style of each independent space. On the black steel board, scarlet western-styled sofas are in good match with suspended steel pipelines. Suspended studio room made of transparent glass windows is a masterful design. The designer artfully and smartly put these contradictory elements together, infuses the space with individualistic aestheticism and an excellent space to inspire the staff to create.

内建築

P94-P95

	1

1. 粗糙的柱壁，灰暗的水泥地面，一目了然的是墙上的三个字“内建筑”。

1. Crude pilasters walls, dark cement floor, and evident “Interior Architecture Design” on the wall.

1	2

1. 裸露的管道比比皆是。
2. 黑漆漆的铁制楼梯张扬地裸露着红色的锈斑。

1. Exposed pipelines can be seen everywhere.
2. Black staircase in ironwork reveals reddish rust stain.

1	2

1．斑驳的砖墙，粗犷而又热情。
2．设计师用厚重的钢铁重组了这个自由空间，将整个工作室设计成了跃层，模糊了传统的空间功能区分，使整幢房子的室内变得更灵活。

1. Stained brick walls are both rough and passionate.
2. The designer used bulky steel to reform this free space, and to design the studio into alternate floor pattern and the traditional functional space section is blurred, which makes the entire house more flexible.

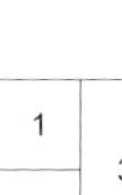

1．别具一格的装饰墙。
2．老仓库的历史底蕴、高大空阔的房屋结构，种种因素结合成创意工作室最理想的办公环境。
3．巨幅的人物肖像照扩张了视觉的深度。

1. Unique styled decorative wall.
2. historical containing, tall and broad structure and other factors together constitute the most ideal office environment of the studio.
3. Huge piece of figure portrait expands the depth of vision.

P102-P103

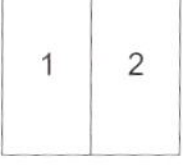

1．纵横交错的钢结构拉伸了整个空间。
2．厚重的钢铁结构与现代的西式沙发搭配得相得益彰。

1. Vertically and horizontally interlaced steel members expand the entire space.
2. Modern western-styled sofas are in good match with bulky steel members.

P104-P105

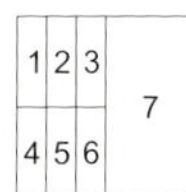

1．悬挂的装饰物。
2．玻璃质地的书报架，强调通透。
3．随处可见的裸露感。
4．独特的装饰物。
5．线条简洁的欧式家具。
6．颇具历史底蕴的小装饰。
7．稳重的木质结构与简洁的线条使中西文化完美结合。

1. Suspending ornaments.
2. Glass shelf for books and newspaper emphasizes penetration.
3. Exposure sensed everywhere.
4. Unique ornaments.
5. European-styled furniture in simple lines.
6. Small ornaments with historical containing.
7. Sedate wood material fitting into compact lines reflects the perfect union of Chinese and occidental cultures.

enter

The Eight Bridge
8号桥

位置：上海市建国中路8－10号 电话：+86－021－64150789

8号桥是一处在1.5万平方米旧工业厂房的基础上建立起来的“时尚创作中心”。改造保留了旧厂房的结构及工业老建筑特有的底蕴，同时又融入了全新的建筑文化理念，从而成为一个激发创意灵感，吸引创意人才的新天地。

名称8号桥，但其实是7座建筑物的连成体，整个园区由四个不同主题的公共空间构成，形态各异，神韵独特。办公楼都由天桥相连，方便各租户之间的沟通与联络。8号桥的建筑风格富于创新，新旧结合的创造，展现独特的魅力风尚。同时整个空间充满了工业文明时代的沧桑韵味。从大门口处法国艺术家创作的大型雕塑《绿门》，到灰砖外墙上鲜亮的玫瑰红色块，以及内部歌剧院般的层叠式休闲吧，都力求创造一个个性化十足、自由度更高，同时又要求能提供方便的办公场所。

在8号桥的设计中，设计师使用了大量的玻璃门窗，通透的玻璃廊房，镶嵌在青灰色的砖墙间，搭配得相得益彰。房顶依然是尖顶瓦面，设计师巧妙设计了矩形玻璃天窗，以保证室内良好的采光。每个办公楼的每个空间都让人感觉高挑、宽敞，布局错落有致。

这儿最独特的是园区设计师留出了很多“租户共享空间”，如商务中心、休闲后街、阳光屋顶等，可以给租户提供许多互动空间，使不同领域的艺术工作者和各类时尚元素在这里互相碰撞，激发灵感和创意。

The Eight Bridge is a vogue creation center built on base of an old industrial plant of an area 15 thousand squared meters. The reconstruction keeps the structure of the old plant building and the special architectural containing, and meanwhile it blends with new architectural and cultural concept. Therefore, it becomes a new world which inspire people to think and create and attract originative talents.

Titled as The Eight Bridge, it is actually a building complex by 7 buildings connected each other. The building complex contains 4 public spaces of different themes, and different forms with different verves. The offices are all linked by overbridges which facilitate the tenants' communication each other. The Eight Bridge is innovative in architectural style, the union of the old and the new shows its unique charm. The entire space is full of the taste of change of the world and time specially belonging to industrially cultural era. The big scaled sculpture work *the Green Gate* at the entrance area, the bright rosy pieces on the brick exterior wall and leisure bar of the inner opera are all individualistic and free enough to create an office space which can provide convenience.

The designer plentifully uses glass doors, windows, and glass porch embedded between blue-grayish brick walls so that they are in perfect match. The roof is furnished with pointed tiles. The designer smartly designs rectangular glass skylight to assure the good lighting on the interior. Every space of every office building seems lofty, spacious and good in layout.

What is the most inimitable is that the designer preserved many “shared spaces by the tenants” such as business center, leisure back street and sunshine roof, etc. which provide the tenants many interactive spaces. Artistic workers in different domains and various voguish elements interplay and collide so that inspirations and originalities are wakened.

P106-P107

1. 8号桥保留了老厂房中那些厚重的砖墙、纵横的管道和斑驳的地面，巧妙地在楼与楼之间用桥连接。同时抛开工业老厂房原有的沉重感，随处可见的是前卫的创意。

1. The Eight Bridge reserves the bulky brick walls, pipelines in length and breadth and stained grounds of the old plants, and smartly uses bridges to link between buildings. Meanwhile, the original ponderosity is scattered and instances of vanguard originality are everywhere.

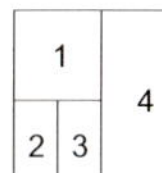

1. 夜景中的8号桥。法国艺术家创作的大型雕塑《绿门》在鹅黄与紫红色的灯光下，温馨而又雅致。

2、3. 灰砖外墙上鲜亮的玫瑰红色块，强调了视觉的冲撞。

4. 每个办公楼的每个空间都让人感觉高挑、宽敞，布局错落有致。

1. The Eight Bridge in night: Large scaled sculpture the Green Gate seems gentle and fragrant under light yellow and purple lights.

2、3. Bright rosy pieces on the gray bricked exterior wall emphasize the visual impact.

4. Every space of every office building makes people feel its tallness, spaciousness and proper-proportion.

P110-P111

1	3
2	

1．色彩丰富、造型奇特的休闲桌椅。
2．通透的玻璃廊房，流露出现代文化的气息。
3．歌剧院般的层叠式休闲吧，错落有致，个性十足。

1. Colorful and fantastically formed leisure desk and chairs.
2. The glass porch expresses the taste of modern culture.
3. Cascaded leisure bar of the inner opera is properly proportioned all highly individualistic.

1	2	3

1．逐级向上的台阶与屋顶纵横的装饰交相辉映。
2．盘旋向上的楼梯。
3．8号桥的设计利用了原厂房的高空间、多层次的布局。

1. Gradually upward steps are reflected by the ceiling.
2. Spiral staircase.
3. The design of the Eight Bridge makes use of the high space, multi-leveled layout of the original plant.

Alibaba.com
阿里巴巴(杭州)

位置：杭州市文三路478号华星时代广场20层　电话：+86－0571－26882688

阿里巴巴作为全球B2B电子商务的著名品牌之一，不仅从事着新科技的产业，整个阿里巴巴总部的设计也让人感觉就像是在贩卖时尚产品的概念店。

阿里巴巴总部由工作区、会议区、经理室、走廊等多种空间组成，通过合理完善的规划，使办公室成为一个具有机能性、时尚前卫，且能发挥高效的工作环境。设计师运用灯光及蓝色的光亮超感隔离墙，营造了科技的场所感。不仅仅带来视觉上的新感觉，而且其表面的光亮、印象派的色彩及虚空间的意境，使之超然于其他风格而具有新现代主义光亮派的特点。

整个办公区域都采用了水泥地面，在灯光的作用下反射着蓝色的光亮，过道上白色的虚线路标更充满了动感。半开放式的工作室采用了中性的灰色调，灰色的隔离墙间隔着透明的玻璃幕墙，玻璃上白色的Alibaba.com自然地体现出蓬勃的企业活力。与办公区截然不同的是，会议区运用的是跳跃的红色，色彩的对比相映成趣，使员工在不同的场合体会到不同的环境和状态。

为了进一步体现本案设计的时尚感，设计师将走廊设计成流线型，配合富有凹凸感的光亮墙面，使空间更具延展性。走廊上抽象的阿拉伯符号及灯光打造了一个虚幻的场景。流动的空间，流动的灵感，流动的科技，透明高效的办公氛围便由此诞生。

As one of the most famous brands of electric business worldwide, Alibaba is not only engaged in scientific industry, but also itself seems like selling products of vogue.

Alibaba.com is composed of various spaces such as working area, meeting area, manager room, corridor and etc. through reasonable planning, the office is turned into an environment which is organically functional, voguish and highly effective. The designer uses lamplight and blue light on the partition wall to create a place full of scientific taste. This not only brings new visual feeling but also its surfacing lucency, impressionist colors and illusive space go beyond other styles.

The entire office area uses cement ground, on which blue light is reflected. The broken line on the ground of aisle gives dynamic sense. Semi-opening office uses neutral grayish color, transparent glass wall stays between grayish partition walls, and “Alibaba.com” on the glass naturally shows the enterprisal vigor. Quite contrary to the office area, the meeting area uses flamboyant red color to make the staff experience a different environment and condition.

In order to express the sense of vogue, the designer designs streamlined corridor, in companion with concavo-convex lucent wall faces to prolong the space. Abstract Arabic numerical expressions and lighting create an illusive scene. The fluid space, fluid inspiration and fluid science and technology, and transparent office environment are born.

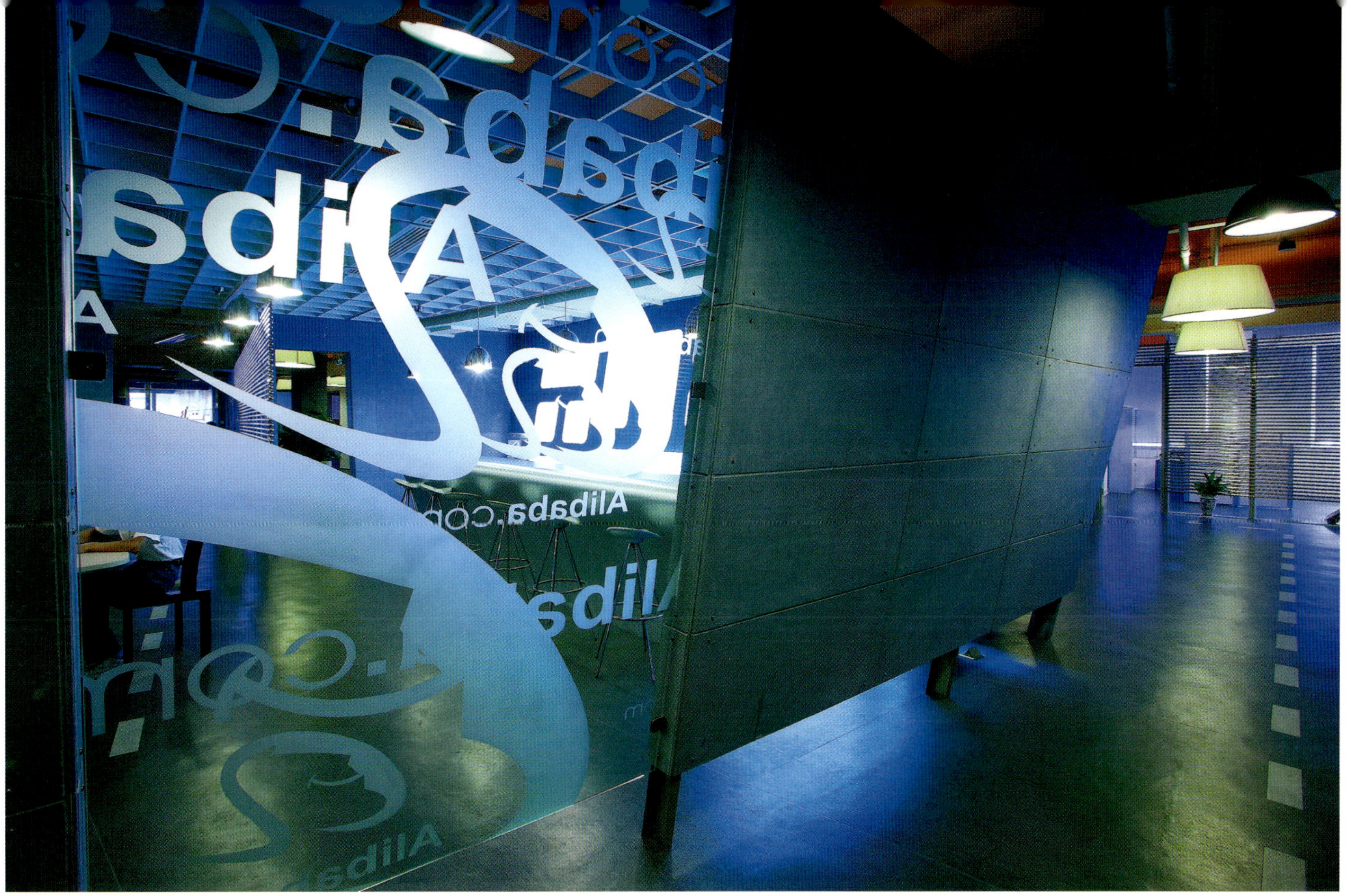

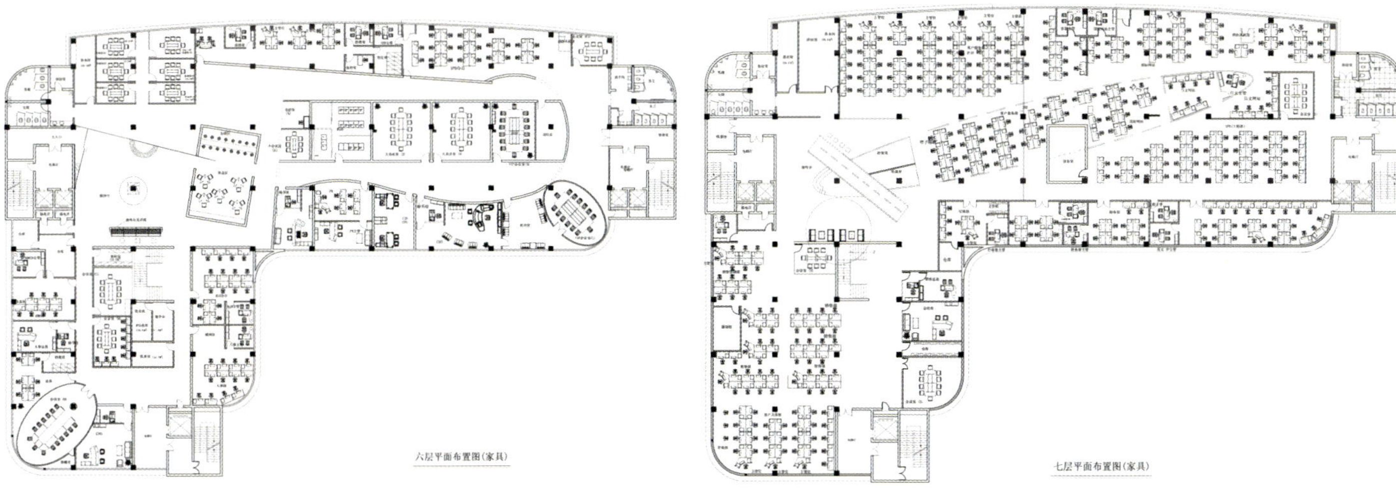

六层平面布置图(家具)

七层平面布置图(家具)

P114-P115

1

1．整个阿里巴巴总部的设计也让人感觉就像是在贩卖时尚产品的概念店。

2. Alibaba seems like concept shop selling voguish products.

1 3
2

1．半开放式的工作室采用了中性的灰色调。
2．与办公区截然不同的是，会议区运用的是跳跃的红色。
3．玻璃上白色的Alibaba.com自然体现出蓬勃的企业活力。

1. Semi-opening office uses neutral grayish color.
2. Quite contrary to the office area, the meeting area uses flamboyant red color.
3. Alibaba.com on the glass naturally shows the enterprisal vigor.

1. 设计师运用灯光及蓝色的光亮超感隔离墙，营造了科技的场所感。
2. 横向延伸的隔离墙加深了空间的层次感。
3. 合金材质弯曲成奇特的流线型长椅。

1. The designer uses lamplight and blue light on the partition wall to create a place full of scientific sense.
2. Horizontally lengthened partition wall increases sense of levels of the space.
3. Alloy is curved to form fancy streamlined bench.

P120-P121

1. 流线型的走廊，配合富有凹凸感的光亮墙面，使空间更具延展性。
2. 走廊上抽象的阿拉伯符号及灯光打造了一个虚幻的场景。
3. 过道上白色的虚线路标更充满了动感。
4. 整个办公区域都采用了水泥地面，在灯光的作用下反射着蓝色的光亮。

1. Streamlined corridor, in companion with concavo-convex lucent wall faces prolongs the space.
2. Abstract Arabic numerical expressions and lighting create an illusive scene along the hallway.
3. White route sign in dotted line on the floor of the passageway seems dynamic.
4. The entire office area bears cement ground, on which blue light is reflected.

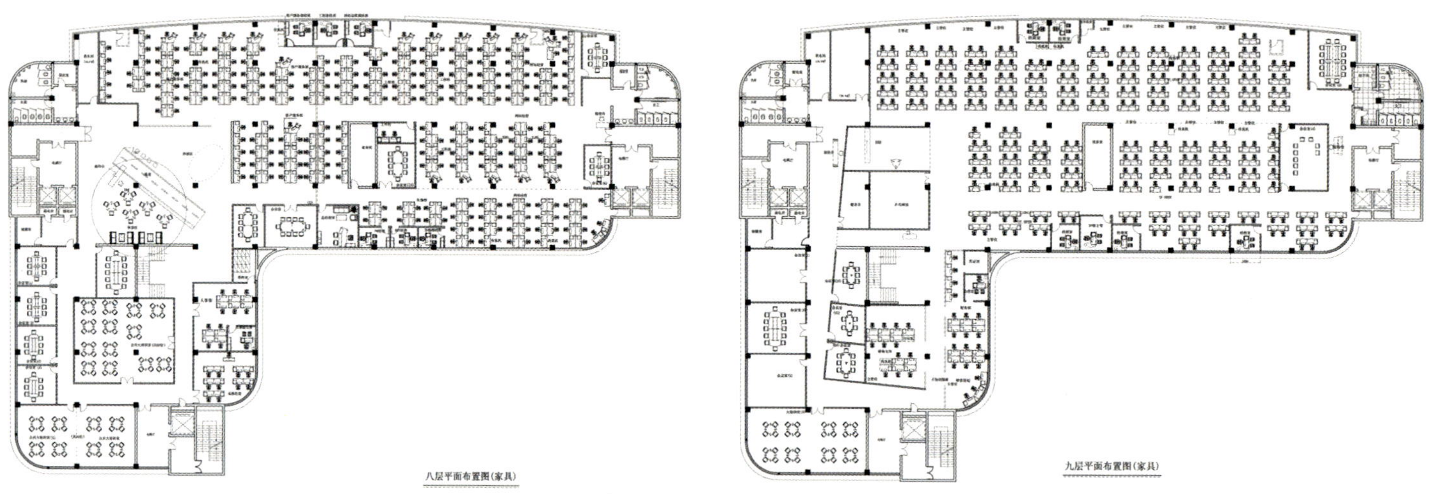
八层平面布置图(家具)
九层平面布置图(家具)

3
2

TaoBao.com
淘宝网

位置：杭州市文三路477号华星科技大厦九层　电话：+86－0571－88155188

淘宝网的工作室强调了e时代办公样式的环境个性，整体建筑布局简约流畅、时尚张扬。设计师将淘宝网的企业文化很好地融入了设计中，无论是色彩丰富的竖向线条、工作室的玻璃幕墙造型，还是充满太空金属感的会议室等，都简洁地表现了网络产业的科技感和现代感。

进入淘宝网的接待大厅，最醒目的就是圆柱形的接待台，黑漆漆的镂孔材质带着优雅的弧度，就好似一艘充满了新奇感的太空船。“淘宝网”和“支付宝”几个大字醒目而又充满活力。为了加强建筑的挺拔感和立体感，整个工作室的墙面及顶部都采用了线条形的橘色、黑色和白色做装饰，丰富而又饱满的色彩点缀在深灰色地面中，产生着跳跃感。富有流动感的造型线条以水平和垂直方向展开，丰富了空间的形象，使工作室充满了运动感和生命力。

淘宝网的工作室被设计成一间间独立的小办公室，靠走廊的门及墙面均采用整幅落地玻璃间墙，透明的玻璃上以简朴的线条密密麻麻地印刻着淘宝网或支付宝的网址，使大幅玻璃墙没有平淡感。每间办公室的设计风格基本一致。白色的墙面上用黑色的粗线条印画着抽象的画面，使得整幅墙面不至于呆板，也凸显了淘宝网活跃的个性。会议室的设计格调基本延续了接待台的风格，半圆形的弧线造型极具智能时代感，体积感强，也延伸了视觉线。

Taobao.com's studio emphasizes environmental individuality of e time office mode, and the entire architectural layout is simple, fluent and voguish. The designer integrates Taobao's enterprisal culture into the design. The colorful vertical lines, studio's glass partition wall forms and metal-like meeting room all expresses the senses of science and technology and modernism of IT industry.

What strikes the eyes most is the column formed reception desk, whose black material has graceful radian as a spacecraft full of novelty. Letters “Taobao.com” and “Zhifubao.com” are eye striking and lively. In order to strengthen the senses of soaring and third dimension of the architecture, the entire studio's wall surfaces adopt orange, black and white in lines as decoration. Sense of jumping occurs in that the ground is interspersed with rich and full colors. Forming lines full of fluid sense outspread in horizontal and vertical directions, which enriches the image of the space and infuses the studio with sense of movement and vitality.

Taobao.com's studio is so designed that it is divided into many independent smaller offices, the door near the corridor and wall all adopt grounded glass partition. Website of Taobao or Zhifubao is thickly printed in simple lines on transparent glass, and there is no plainness of the big piece of glass wall. The design of every office room is the same by and large. Abstract pictures are printed in bold black lines on the white wall, which makes the whole piece of wall avoid dullness and gives prominence to the active character of Taobao.com. the design of meeting room succeeds to the style of reception desk. Its semi-circled arced form is full of intelligence-associated sense of era, sense of cubage and of visual prolongation.

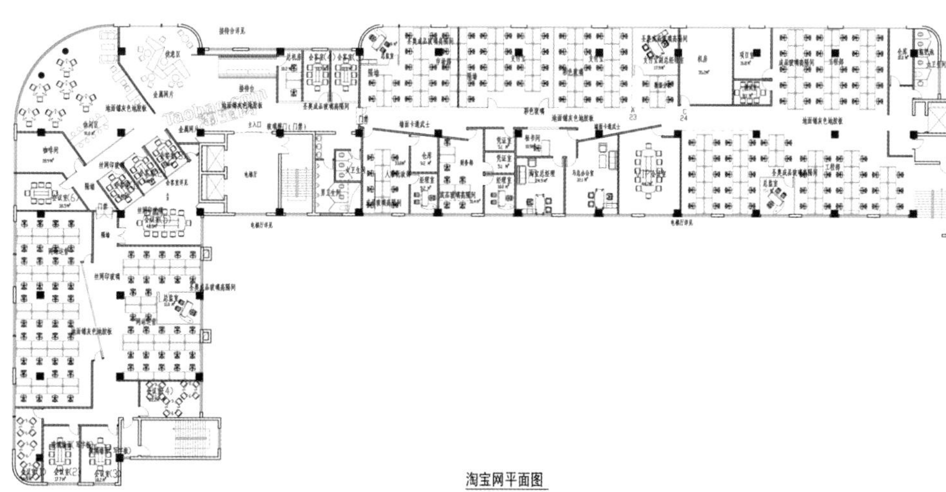

淘宝网平面图

1

1．透明的玻璃上以简朴的线条密密麻麻地印刻着淘宝网的网址，使大幅玻璃墙没有平淡感。

1. Website Taobao or Zhifubao is thickly printed in simple lines on transparent glass, and there is no plainness of the big piece of glass wall.

Taobao.com

1	2

1．白色的墙面上用黑色的粗线条印画着抽象的画面，使得整幅墙面不至于呆板，也凸显了淘宝网活跃的个性。
2．设计师将淘宝网的企业文化很好地融入了设计中。

1. Abstract pictures are printed in bold black lines on the white wall, which makes the whole piece of wall avoid dullness and gives prominence to the active character of Taobao.com.
2. The designer integrates Taobao's enterprisal culture into the design.

1	2

1. “淘宝网”和“支付宝”几个大字醒目而又充满活力。
2. 黑漆漆的镂孔材质带着优雅的弧度，就好似一艘充满了新奇感的太空船。

1. Letters “Taobao.com” and “Zhifubao.com” are eye striking and lively.
2. Black bored material has graceful radian as a spacecraft full of novelty.

淘宝网

1	2	4
	3	

1. 半圆形的弧线造型极具智能时代感，体积感强，也延伸了视觉线。
2. 整个工作室的顶部都采用了线条形的橘色、黑色和白色做装饰，极富立体感。
3. 丰富而又饱满的色彩点缀在深灰色地面中，产生跳跃感。
4. 富有流动感的造型线条以水平和垂直方向展开，丰富了空间的形象，使工作室充满了运动感和生命力。

1. Semi-circle arcs form is extremely full of sense of intelligence times, sense of third dimension and prolong the line of sight as well.
2. The entire studio's ceiling adopts orange, black and white in lines as decoration, and it is full of sense of third dimension.
3. Sense of jumping occurs in that the ground is interspersed with rich and full colors.
4. Forming lines full of fluid sense outspread in horizontal and vertical directions, which enriches the image of the space and infuses the studio with sense of movement and vitality.

Yahoo.com.cn
雅虎中国

位置：北京市光华东路甲8号和乔大厦B座6层　电话：+86－010－65833721

雅虎中国作为网络产业的门户网站，在强调尊重人性、尊重创造的前提下，千方百计改善和创造更好的工作与生活氛围，从而造就出一种更为高效和谐、愉悦的工作环境。设计师在构成设计上运用了多边形和长方形两种几何图形在空间相互穿插，造型手法极具现代感，力图通过几何图形的变化和轴线的灵活运用，创造出一个具有独特肌理的创意空间。同时，通过建筑空间尺度的塑造，以半开放式的空间，在不同围合空间的工作室里，采用不同的设计来体现不同的主题。

雅虎中国工作室采用的是自由灵动的格局，大量运用了弧形曲线元素，使内部平面结构整体呈不规则形状。整个办公区除开放式的会议休闲区之外，其他办公室和会议室都被设计成灵气四动、长方体的几何体。设计师充分利用现代建材、结构，表现简洁、通透，用鲜艳的橘红色、张扬的阿拉伯数字来寻求整个构图的韵律与灵动性，用圆滑的弧度表现出严肃的几何图形。独具特色的造型，令这个空间显得明快而又活跃。为了更好地搭配几何体的设计，工作室的照明设施也被设计成了长方体，随意地斜跨在房顶上。

工作室吊顶的设计也随着空间形式、办公家具的改变而变化多端。时而是错综复杂的管道纵横交错，时而又是中空的造型，在强调层次感的同时，更显简洁、明快。

雅虎中国无论从空间上，还是色彩上，都表达了时尚感、跳跃感，充满想象力及活力。

As one of representative websites of Chinese IT industry, Yahoo.com.cn tries every means to improve and create better working and living atmosphere so that a more efficient, harmonious and pleasing working environment is created. In terms of composing design, the designer uses interlaced polygons and rectangles in the space; and he tries to create a space of unique texture and originality through the variation of the geometrical forms of modern taste. Meanwhile, through shaping the architectural space dimension, different designs are to express different themes with respect to different enclosed studio rooms.

The studio of Yahoo adopts free and flexible layout, which plentifully uses arced curved elements which makes the interior plane structure irregular. Except for the leisure area, the other offices within the whole office are all designed as flexible cuboidal form. The designer fully utilizes modern building material, structure to express simplicity and sense of being through; uses fresh orange red and Arabic numbers to seek rhythm and flexibility of the entire composition of the picture; and uses smooth radian to express serious geometrical patterns. The unique forms make the space both bright and active. In order to make a good match with the geometrical patterns, the lighting facilities are all designed as rectangular forms and they casually dwell under the ceiling.

The design of hanging ceiling of the studio differs according to the variation of the form of the space and of the office furniture. Some of them are as complicated pipelines and some are hollow in form.

Yahoo China expresses sense of vogue and jumping no matter in terms of space or in terms of color, and it is full of imagination and liveliness.

P130-P131

	1

1．鲜艳的橘红色表达了时尚感、跳跃感。

1. Fresh jacinth expresses the sense of vogue and jumping.

1	2	3

1．多边形的几何体显得新颖而又时尚。
2．吊顶处纵横交错的管道，强调了层次感。
3．墙上的柜面装饰夸张的阿拉伯数字，充满了活力。

1. Polygonal geometrical form seems both novel and voguish.
2. Complicatedly intervened pipelines at hanging ceiling emphasize the sense of levels of space.
3. Exaggerative Arabic numbers are full of vigor.

1. 设计师在构成设计上运用了多边形和长方形两种几何图形在空间相互穿插，造型手法极具现代感。
2. 工作室采用自由灵动的格局，大量运用了弧形曲线元素，使内部平面结构整体呈不规则形状。
3. 走廊上整幅的落地玻璃幕墙，采光通透。

1. In terms of composing design, the designer uses interlaced polygons and rectangles in the space; and the forms bear modern taste.
2. The studio of Yahoo adopts free and flexible layout, which plentifully uses arced curved elements which makes the interior plane structure irregular.
3. The whole piece of grounded glass partition wall does well in light introduction.

等候区
接待区
VP(一)
VP(二)
休息间
总裁办公室
大会议室(一)
女厕
男厕
设备
小间
电气
小间
电梯厅
接待台
通道
通道
总文印
茶水
行政(7+1)
法务(5)
Marketing(31人)
文件档案储藏室
文件档案储藏室
水房
空调机房
电梯厅
IT机房
法务总监
人力资源 (15+2)
财务 (22)
财务总监
电气
小间

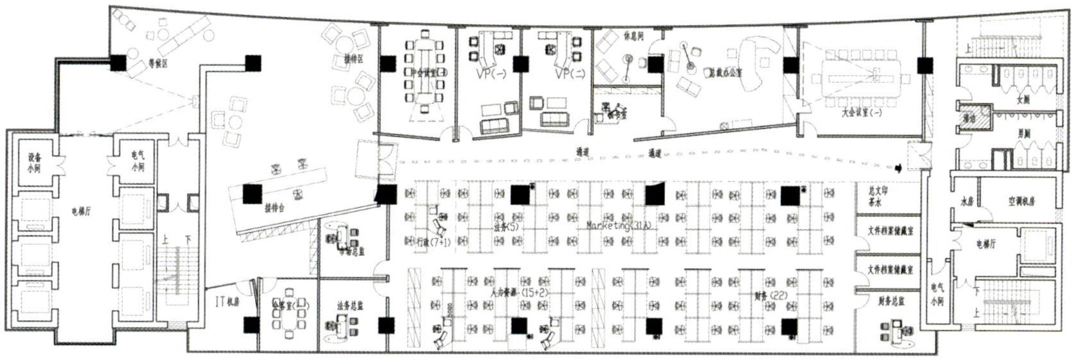

九层平面图

1	2	3

1．张扬的阿拉伯数字强调整个构图的韵律。
2．设计师充分利用现代建材、结构，表现简洁、通透。
3．整个办公区除开放式的会议休闲区之外，其他办公室和会议室都被设计成灵气四动、长方体的几何体。

1. Exaggerative Arabic numbers emphasize the rhythm of the entire picture composition.
2. The designer fully utilizes modern building material, structure to express simplicity and sense of penetration.
3. Except for the leisure area, the other offices within the whole office are all designed as flexible cuboidal form.

设备
小间
电气
小间
电梯厅
43.100
上
下
小会议室（八）
IT机房
员工健身区
中会议室（四）
技术部总监二
女厕
男厕
水房
空调机房
电梯厅
电气
小间
下
上

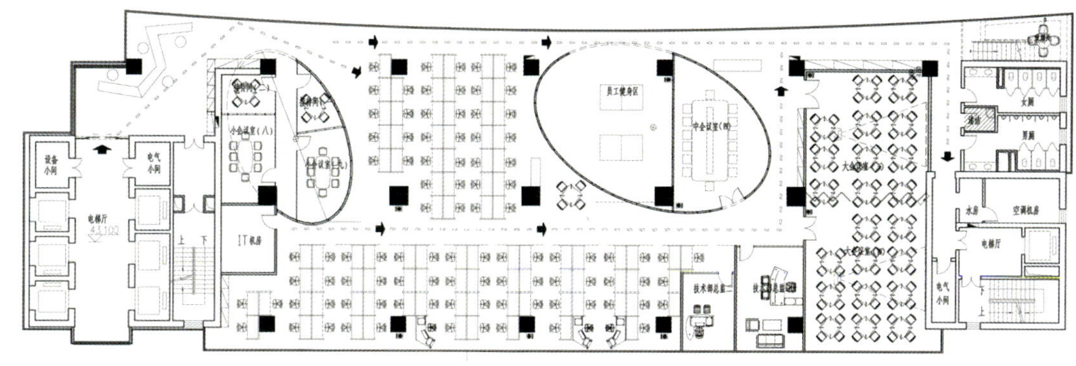

十五层平面图

hoo.com.cn

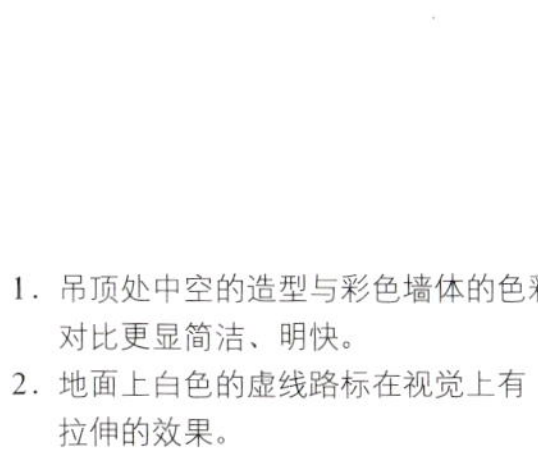

1. 吊顶处中空的造型与彩色墙体的色彩对比更显简洁、明快。
2. 地面上白色的虚线路标在视觉上有拉伸的效果。

1. The hollow form at the hanging ceiling seems simple and brisk.
2. White route sign in dotted line on the floor has an effect of prolonging visually.

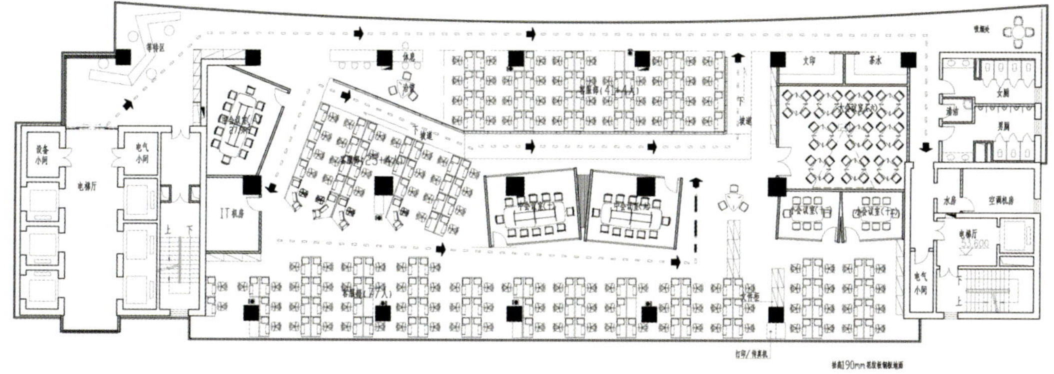

十七层平面图

1. 设计师力图通过几何图形的变化和轴线的灵活运用，创造出一个具有独特肌理的创意空间。
2. 鲜艳的橘红色搭配夸张的阿拉伯数字，凸显灵动性。
3. 工作室的照明设施也被设计成了长方体，随意地斜跨在房顶上。

1. The designer tries to create a space of unique texture and originality through the variation of the geometrical forms and flexible use of axis.
2. The collocation of fresh jacinth and exaggerated Arabic numbers shows cleverness.
3. The lighting facilities are all designed as rectangular forms and they casually dwell under the ceiling.

1	2	5
3	4	

1. 用玻璃形成的隔断强调了通透性。
2. 带箭头的路标令工作室充满想象力。
3. 用圆滑的弧度表现出严肃的几何图形。
4. 地面上昆虫的造型可爱而有趣，为创意生活增添了灵感。
5. 独具特色的造型，令这个空间显得明快而又活跃。

1. Glass partition enhances the sense of penetration.
2. Arrowed route sign makes the studio full of imagination.
3. Smooth arc expresses sober geometrical forms.
4. The forms of insects on the ground are lovely and interesting, which inspire the originality of life.
5. Unique forms make the space brisk and bright.

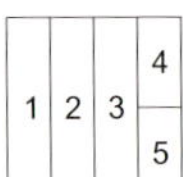

1、2. 设计师通过建筑空间尺度的塑造，以半开放式的空间，
在不同围合空间的工作室里，采用不同的设计来体现
不同的主题。
3. 简约的设计风格。
4、5. 黑白色调的办公家具，简洁大方。

1、2. Through shaping the architectural space dimension, different designs are to express different themes with respect to different enclosed studio rooms.
3. Concise design style.
4、5. Office furniture in black and white seems simple and generous.

China Greentown Architectural Design
绿城建筑设计公司

位置：杭州市西湖区玉古路161号 电话：+86－0571－87995338

绿城建筑设计公司的设计主调融合了中西方的不同风格，在西方的现代简约中注入了东方古朴传统的美，注重意境的渲染。整体的建筑设计讲究线角的丰富细腻，讲究形体变化的韵律，讲究窗、门、楼梯等建筑构件组合的比例关系，让建筑在转折进退、高低错落中增强光影变化，增加细节的层次，以此来丰富建筑本身的语言，给人一种自然而富有柔韧性的美感。使得整个办公楼不再单调而呆板，充满了灵性。

推门而入，便感受到了跃层空间的敞亮，设计师利用高度上的局部变化营造错落、丰富的屋顶轮廓和天际线，结合局部架空的手法丰富整个建筑造型。建筑立面采用大面积玻璃幕墙处理，外观现代、通透，充满了时尚质感。内墙中赫然耸立着一堵色彩参差不齐的黄褐色砖墙，在周围玻璃幕墙的映衬下显得格外瞩目。设计师在保留城市现代气质的同时更注重了文化意境的塑造，给人以更多的审美空间和趣味。

造型独特的楼梯通过各种金属结构将玻璃等连接在一起，形成柔顺、通透的的楼道。顶层的设计采用了歌剧院式建筑布局，依自然的坡度而建，疏密有致。观景屋顶引入更多的户外景色与阳光，同时使得室内的光影变化更为丰富。工作室采用了自由开放的空间设计，在金属、玻璃等现代建筑的元素中添加了古色古香的仿红木桌椅及原木色的书桌，文化的内涵并不直白地表露出来，而是凝敛于形体之内， 显得端庄典雅、气质高贵。

The theme of China Greentown Architectural Design combines the oriental and occidental styles. It integrates oriental primitive beauty into occidental modern simplicity and pay attention to the rendering of artistic conception. The entire design of the architecture is in pursuit of richness and subtlety of the lines, rhythm of change in forms, and proportion among windows, doors and staircases to enrich the architecture itself. It gives people a kind of beauty full of nature and flexibility which makes the entire office building not monotonous any more.

Entering the inside, one can feel the spaciousness and brightness of the alternate storied space. The designer utilizes the local change in height to build random, ample-formed roofs and skyline, and enrich the entire building form through locally building on stilts. The facade adopts massive pieces of glass curtain walls that seem modern, transparent and full of sense of vogue. A piece of yellow-brown bricked wall stands still in the middle of the walls on the interior, and it is very striking in the surrounding of glass walls. The designer pays much attention to cultural artistic conception as well as modern city's identity. This leaves more space of aestheticism and interests.

Uniquely formed staircases are connected each other by various metal structures therefore smooth and fluent stairways are formed. The design of top floor adopts the architectural layout of an opera; it is built based on natural slope with certain spacing. View-taking roof introduces more outdoor scenes and sunshine and meanwhile enriches the interior with the variation of the lights and shadows. The studio adopts free opening space design, which adds the antique-tasted mahogany-mimic tables and chairs and original wood colored desk, through which the cultural containing is so indirectly expressed that it seems sublime, elegant and noble.

浙江绿城建
GREENTONARCHITECTURALDE

P146-P147

	1

1．挑高的大厅中灰色的隔离墙极具大气。

1. The partition walls in the high rising hall seems grand.

1	2

1．造型独特的楼梯通过各种金属结构将玻璃等连接在一起，形成柔顺、通透的的楼道。
2．建筑立面采用大面积玻璃幕墙处理，外观现代、通透，充满了时尚质感。

1. Uniquely formed staircases are connected each other by various metal structures therefore smooth and fluent stairways are formed.
2. The facade adopts massive pieces of glass curtain walls that seem modern, transparent and full of sense of vogue.

还书处

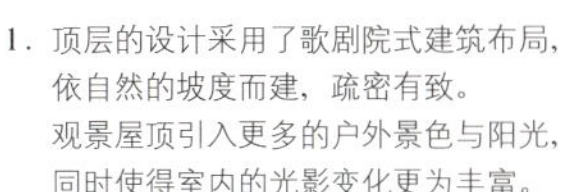

1. 顶层的设计采用了歌剧院式建筑布局，依自然的坡度而建，疏密有致。观景屋顶引入更多的户外景色与阳光，同时使得室内的光影变化更为丰富。
2. 内墙中赫然耸立着一堵色彩参差不齐的黄褐色砖墙，在周围玻璃幕墙的映衬下显得格外瞩目。

1. The design of top floor adopts the architectural layout of an opera; it is built based on natural slope with certain spacing. View-taking roof introduces more outdoor scenes and sunshine and meanwhile enriches the interior with the variation of the lights and shadows.
2. A piece of yellow-brown bricked wall stands still in the middle of the walls on the interior, and it is very striking in the surrounding of glass walls.

1	2	5
3	4	

1．建筑高低错落中增强光影变化，增加细节的层次，以此来丰富建筑本身的语言。
2．利用高度上的局部变化营造错落。
3．设计师使用了大量的表现城市现代气质的元素。
4．整体的建筑设计讲究线角的丰富细腻。
5．工作室采用了自由开放的空间设计。

1. The entire design of the architecture is in pursuit of richness and subtlety of the lines, rhythm of change. in forms, and proportion among windows, doors and staircases to enrich the architecture itself.
2. The designer utilizes the local change in height to build random, ample-formed roofs and skyline.
3. The designer uses many elements that can express modern urban temperament.
4. The entire architectural design is in pursuit of rich and detailed lines.
5. The studio adopts free opening space design.

1．现代简约的楼梯设计。
2．设计师在保留城市现代气质的同时更注重了文化意境的塑造，给人以更多的审美空间和趣味。
3、4．在金属、玻璃等现代建筑的元素中添加了原木色的书桌，文化的内涵并不直白地表露出来，而是凝敛于形体之内，显得端庄典雅、气质高贵。

1. Modern simple design of staircase.
2. The designer pays much attention to cultural artistic conception as well as modern city's identity. This leaves more space of aestheticism and interests.
3、4. The antique-tasted mahogany-mimic tables and chairs and original wood colored desk are introduced into the elements of metals, glass and other modern ones, through which the cultural containing is so indirectly expressed: sublime, elegant and noble.

Wise Windows
智汇堂

位置：杭州市拱墅区杭印路49号 电话：+86－0571－88966706

智汇堂给人带来的感觉就是现代、简洁、明快。整个工作室的空间分配相当简洁，少量的点睛之品营造出简洁宁静的工作环境。环境布置和配饰的选择绝没有被过多思想影响和左右，空间中的装饰线条最终都被归为两类－水平和垂直。在这里，色彩的作用不是突出某件事物，而是与其他元素相互和谐地结合。同时在材料上的选择也是遵循这样的思路，材料分类并不多，皆在体现现代、明快的个性。设计师非常巧妙地利用了玻璃的通透、白色的明亮以及线条的延伸，强调了工作室干净、利落、自然的个性，使原本空间不大的智汇堂显得宽敞明亮。

白色是整个工作室的主色调，白色的墙、白色的顶、白色的桌椅、白色的地板，整个空间一气呵成，纯净的白色在视觉上大大地起到了扩大面积的作用，使整个空间显得整洁宽敞。背景墙采用的同样是白色的竖条纹，层层递进的设计延伸了空间，也开阔了视野。设计师在最大化拉伸空间的同时，将白色与玻璃幕墙相融合，让整个空间在白色中拉开层次。大面积轻盈通透的玻璃幕墙与白色的墙体虚实互补，造型浑然一体，将明快的空间特征表现得淋漓尽致。

整个工作室布置得相当简洁。独具北欧风情的办公家具简约而又明亮，外观线条流畅，外形隽颖瑰丽，勾勒出建筑的时尚与现代。

What impress the people by the Wise Windows are modernism, simplicity and clear brightness. The entire studio space is less occupied, a few focal article contribute to a simple and quiet working environment. The environment and decoration are not dominated by much thoughts and ideas, and the decorative lines are eventually classified into two: vertical and horizontal ones. In here, the role of color is not to emphasize some certain thing but to become a union with the other elements. Meanwhile, in terms of choosing material such idea is followed. There is less kinds of materials but all of them try to express modern and bright individuality. The designer smartly utilizes the transparency of glass, brightness of white color and prolongation of lines to emphasize cleanly, tidy and natural characters and turns the normal space of the Wise Windows into spacious and bright one.

White is the dominating color of the entire studio, the white wall, ceiling, chairs and floor all seems in a unity. The pure white plays a role of extending the space area visually. The background wall uses the vertical stripes in same white color, and the design in such way prolongs the space and broadens the vision as well. The designer maximum the space and meanwhile integrates the white color into glass partition wall to make the entire space arranged in levels in the white color. Transparent glass partition walls in large area are in contrast with white wall blocks, which form a perfect unity and express the bright character of the space wholly.

The entire studio is simply furnished. Nordic styled office furniture is simple, bright, pretty and fluent in outline, which express the taste of modernism and vogue.

1．背景墙采用的是白色的竖条纹，层层递进的设计
延伸了空间，也开阔了视野。

1. The background wall uses the vertical stripes in same white color, and the design in such way prolongs the space and broadens the vision as well.

Creative design

1	2	4
3		

1．大面积轻盈通透的玻璃幕墙与白色的墙体虚实互补，造型浑然一体，将明快的空间特征表现得淋漓尽致。
2．独具北欧风情的办公家具简约而又明亮，外观线条流畅，外形隽颖瑰丽，勾勒出建筑的时尚与现代。
3．纯净的白色在视觉上大大地起到了扩大面积的作用，使整个空间显得整洁宽敞。
4．设计师为了让白色不至于有闷涩的感觉，在局部大胆运用了鲜亮的色块，营造欢快活泼的空间环境。

1. Transparent glass partition walls in large area are in contrast with white wall blocks, which form a perfect unity and express the bright character of the space wholly.
2. Nordic styled office furniture is simple, bright, pretty and fluent in outline, which express the taste of modernism and vogue.
3. The pure white plays a role of extending the space area visually and makes the entire space cleanly and spacious.
4. In order not to be bored by white color, the designer boldly uses fresh and bright colors in parts to built buoyant and joyous space.

CA-Group
文筑国际

位置：上海市大连路970号海上海新城9号楼706室　电话：+86－021－33773003

文筑国际是一个典型的小型工作室。设计师合理布局了每个空间，充分利用了工作室的高度，把靠右边的空间划分为两个层面，这样一来，除底层的共享空间外，还形成了一个悬在半空中的工作空间。同时将工作室的特性与功能融入到设计中，整体设计风格简单、随意又富有文化气息。

步入文筑国际，最突出的感觉是不像是一个传统意义上的工作室，而更像是一个空间紧密、布局合理的图书馆阅览室。整个工作室最引人注目的就是左右两排“顶天立地”的格子书橱，30cm见方的格子内放置了各式各样、色彩缤纷的书籍，色彩分明，干净利索，既丰富了空间，又充当了墙面。因地制宜制作的书柜，恰到好处地填补了整面白墙带来的单调感。宽松的工作空间、白色的墙、白色的桌椅，再配以办公桌前大幅的玻璃窗，使得工作环境更显宽敞明亮。

文筑国际的一层被设计成一个开放式的办公区，设计师将文化与现代节奏有机结合，使得空间产生品位与简约相统一。在一楼通往二楼的楼梯处，设计了一堵银灰色的金属隔离墙，隔离墙除了达到楼梯本身连接上下层的作用，也造成了视觉上的冲击力，并起到了划分不同功能区的作用。二楼是一个半隔离的会议室，由于是搭建的“空中楼阁”，因而空间显得较狭窄。设计师为了扬长避短，巧妙地将吊顶设计成透明的方格状，既充当照明，又可作装饰，为整个工作室带来了明快的生气。

CA-Group is a traditional small-scaled studio. The designer reasonably arranges every space of it and fully utilizes the height of it. The space on the right is divided into two levels so that there is a hanging working space in addition to the bottom share space. Meanwhile, the character and functions are integrated into the design which is generally simple and full of cultural sense.

Within CA-Group, what people feel the strongest is that it doesn't look like a studio in traditional sense, but like a space-compacted and well arranged reading room of a library. The most eye-striking in the entire studio is the ceiling reaching book cabinet which justly fills up the whole piece of white wall and the monotony it brings forth. Loose working space, white walls and white desks and chairs in companion with large pieces of glass window before the desks make the working environment more spacious and bright.

The first floor of CA-Group is designed as a open office area, in which the designer organically combines something cultural and modern rhythm so as to unify the taste of the space and simplicity. At where the first floor staircase leading to the second floor, a piece of silver grayish metal partition wall is set, and it serves as both a member which parts different functional areas and a member that is a visual impact. At second floor is a semi-isolated meeting chamber hanging in the air, and due to this, the space seems comparatively narrow. In order to exert advantage and avoid shortage, the suspension ceiling is smartly designed as transparent latticed forms, which can be used both as lighting and ornaments and brings bright liveliness to the entire studio.

TADAO ANDO

P160-P161

	1

1. 文筑国际更像是一个空间紧密、布局合理的图书馆阅览室。

1. CA-Group is more like a space-compacted and well arranged reading room of a library.

1	2

1. 一层被设计成一个开放式的办公区，设计师将文化与现代节奏有机结合，使得空间产生品位与简约相统一。
2. 设计师合理布局了每个空间，充分利用了工作室的高度，把靠右边的空间划分为两个层面。

1. The first floor of CA-Group is designed as a open office area, in which the designer organically combines something cultural and modern rhythm so as to unify the taste of the space and simplicity.
2. The designer reasonably arranges every space of it and fully utilizes the height of it. The space on the right is divided into two levels.

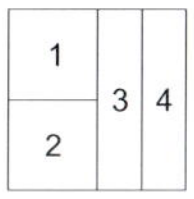

1．在一楼通往二楼的楼梯处，设计了一堵银灰色的金属隔离墙。
2．线条简洁流畅的内部设计。
3．二楼半隔离的会议室。
4．设计师为了扬长避短，巧妙地将吊顶设计成透明的方格状，既充当照明，又可作装饰，为整个工作室带来了明快的生气。

1. At where the first floor staircase leading to the second floor, a piece of silver grayish metal partition wall is set.
2. Fluent, simply lined interior design.
3. Semi-isolated meeting chamber at second floor.
4. In order to exert advantage and avoid shortage, the suspension ceiling is smartly designed as transparent latticed forms, which can be used both as lighting and ornaments and brings bright liveliness to the entire studio.

1	2

1．空中搭建的“空中楼阁”。
2．工作室内放置了各式各样、色彩缤纷的书籍，色彩分明，干净利索。

1. Chamber hanging in the air.
2. Colorful books are placed in bookshelves in the studio.

1	2

1．左右两排“顶天立地”的格子书橱既丰富了空间，又充当了墙面。
2．宽松的工作空间，白色的墙，白色的桌椅，使得工作环境更显宽敞明亮。

1. The ceiling reaching book cabinet on left and right both enrich the space and serve as wall.
2. Loose working space, white walls and white desks and chairs in companion with large pieces of glass window before the desks make the working environment more spacious and bright.

Jian Space
建境

位置：上海市闸北江场三路309号　电话：+86－021－65790747

建境的设计师在设计中大胆调动古典与现代两种相对立的元素，让它们在既自由又规整，既夸张又严谨的氛围中，充分碰撞。整个空间以传统风格和现代实用为依托，处处显露着形似单纯却折射着东方特有的稳重沉着，让整个空间沉浸于东方情结洗礼中。

工作室的空间运用最基本的几何形状，简约至极，塑造出简洁、干练、大气。开放式的办公空间设计现代、简洁、新颖。大面积木制办公家具与胡桃木色的内装饰，怀有东方色彩。设计师利用色彩、线条、玻璃等元素，完成了水平方向的从公共空间向半公共空间的过渡。

建境作为一个复式的结构，充分发挥了楼梯的装饰作用。以高科技派常用的光亮的现代建材表述文脉，这使工作室的设计呈现出一种折中的风采，也成为设计的一个亮点。楼梯处的墙体采用木质的材料，外观质朴大气。淡褐色的色彩本身给人一种华丽的感觉，设计师又用夸张的手法制造出纵横的线条感，与大面积的金属幕墙相结合，从设计细节上体现出横竖结合、凹凸有致的特点，表现手法非常现代。使楼梯从外观看起来稳重又不失活跃，而且颇具档次。同时，引入高大植物、芬芳花草，将自然景色很好地融于建筑物之中，充分表达了人对自然的向往。绿色植物在金属的映衬下，洋溢着天然与时尚的尊贵气息。

The designer of Jian Space boldly exerts two contradictive elements classical and modern ones. They are allowed in the free yet orderly, exaggerated yet solemn atmosphere to collide with each other. The entire space is based on traditional style and modern practicability. Meanwhile it shows formally simplified yet steady and deep meanings from the orient, and set the whole space in the oriental complex.

The studio space uses the geometrical forms to realize the senses of being simple, tidy and generous. The design of open office space is apt to modern, cleanly, simple and novel. Massive use of wooden office furniture and walnut-colored inner decoration really infuse it with oriental tastes. The designer utilizes colors, stripes, glass and etc. to complete the transition from public space to semi-public space in horizontal direction.

Being a structure complex, Jian Space fully exerts the decorative role of staircase. Bright modern architectural materials which are usually adopted by high-tech style are used to express cultural successions. This makes the studio more of a style full of eclecticism and a highlight of the design. the wall next to the staircase adopts wooden material which is rustic and simple in appearance. Light brown itself gives people a sense of being luxurious, and the designer creates sense of vertical and horizontal stripes through exaggerative manner. In companion with metal curtain wall, the design expresses the characteristic of being both vertical-horizontal and concavo-convex, which is so modern that the staircase seems both steady and lively seen on the exterior. Meanwhile, tall plants and fragrant flora and herbs are introduced to make the nature melted into architecture and people's love of nature is fully expressed. Set off by the metal members, the green plants' nobility overflows.

1

1．设计师利用色彩、线条、玻璃等元素，完成了水平方向的从公共空间向半公共空间的过渡。

1. The designer uses colors, stripes, glass and other elements to horizontally complete the transition between public space and semi-public space.

1	2

1. 淡褐色的纵横线条与大面积的金属幕墙相结合，从设计细节上体现出横竖结合、凹凸有致的特点。
2. 以高科技派常用的光亮的现代建材表述文脉，这使工作室的设计呈现出一种折中的风采，也成为设计的一个亮点。

1. Vertical and horizontal stripes in light brown are so in companion with massive metal curtain wall that the design expresses the characteristic of being both vertical-horizontal and concavo-convex.
2. Bright modern architectural materials adopted by high-tech style are used to express cultural successions. This makes the studio more of a style full of eclecticism and a highlight of the design.

Design Ideas USA
美国DI设计库

位置：杭州市拱墅区杭印路49号6号楼 电话：+86－0571－88238816

美国DI设计库是现代与传统激情演绎的成果。传统的红木色家具流淌着浓郁的传统色彩，工字钢和玻璃、田园气息的藤椅和金属材质的对比，形成了传统元素与现代时尚的空间对话。空间总体设计形式简练，大方的结构和丰富的细节处理，结合黑色的钢材，让人觉得这是一个对空间设计美学有创意追求的场所。

一踏进美国DI设计库，宽敞的空间、挑高的大厅、斑驳的地面，第一时间就让人联想到这是一个由老仓库而成的工作室。老厂房原来的柱子和钢结构都被保留了下来，但它们都已经穿上了时尚的“外衣”。原本厂房的内部空间非常大，因此设计师利用了层高的优势，应用了多层次的空间设计理念。“悬空”在两个不同水平面的走道、由地面向上延伸的工字钢，无论是在垂直线条还是在水平线条，均设计了丰富的层次，强调了空间的精髓。圆形的接待台富有创意，又具有动感，体现了现代的时尚感。红木色的办公家具、厚重的台面不仅能体现出使用者的尊崇品位，同时折射着东方特有的稳重沉着。颇具东方风韵的装饰物也为整个空间增添了不少传统的色彩。

美国DI设计库的整个装饰颜色主要以黑色的铁架、透明的大幅玻璃、大红色的置物架、原木色的桌椅构成，既有现代主义应有的热情，又不乏古典主义专有的理性。

Design Ideas USA is the fruit of modernism and tradition. Traditional mahogany furniture expresses thick colors of tradition. “I” steel member against glass and cane chairs against metal texture all form the dialogue between traditional elements and modern voguish space. The design of the entire space is simple and tidy in style. Generous structure and rich treatment of details in combination of black steel make people consider it a place where there is a pursuit of aestheticism and originality of space design.

On the interior, the capacious space, high rising hall and stained ground all make people associate it with an old warehouse in the first place. The columns and steel structures of the old plant are reserved, but they are now dressed in coat of vogue. Originally, the interior space of the old plant is very large so that the designer utilizes the advantage of its multistory and develops the design concept of multi-spaces. Stretching upward from the ground, the “I” steel members are suspended above two walkways in two different planes. No matter in verticality lines or horizontality, the essence of space is interpreted abundantly. The circular reception desk is originative and full of sense of movement and expresses modern vogue. Mahogany colored office furniture and bulky desktop not only reveal the user's dignity and meantime they reflect oriental steadiness and equanimity. The ornaments with oriental taste add much traditional colors to the entire space.

The entire decoration is composed by the colors of the black iron racks, transparent glass in large pieces, red shelves and original wood-colored chairs, which have the passion of modernism and rationality of classicism.

Design Ideas

P174-P175

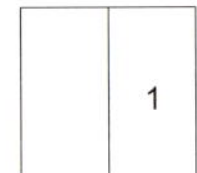

1．圆形的接待台富有创意，又具有动感，体现了现代的时尚感。

1. The circular reception desk is originative and full of sense of movement and expresses modern vogue.

1	2	3

1．排列的灯柱，极富韵律感。
2．厚重的台面折射着东方特有的稳重沉着。
3．设计师利用了层高的优势，应用了多层次的空间设计理念。

1. Lamp colonnade is rich in rhythm.
2. The bulky desktop reflects oriental steadiness and equanimity.
3. The designer utilizes the advantage of its multistory and develops the design concept of multi-spaces.

1．由地面向上延伸的工字钢，丰富了层次，强调了空间的精髓。
2．半透明的装饰隔离墙富有现代感。
3．大红色的家具简洁明快。

1. Stretching upward from the ground, the "I" steel members enrich the levels and emphasize the essence of space.
2. Semi-transparent partition wall is full of modernism.
3. Bright red furniture seems brisk and concise.

1 2 3

1. 颇具东方风韵的装饰物也为整个空间增添了不少传统的色彩。
2. 纵横交错的房顶设计，加深了层次感。
3. 宽敞的空间、挑高的大厅、斑驳的地面，第一时间就让人联想到这是一个由老仓库而成的工作室。

1. The ornaments with oriental taste add much traditional colors to the entire space.
2. Ceiling is designed in length and breadth, which enhance the sense of levels.
3. The capacious space, high rising hall and stained ground all make people associate the very studio with an old warehouse in the first place.

LOFT
49
拱墅區第三屆運河文化藝術節
杭州市拱墅區新型文化產業區建設研討會
FREE SPACE
2004.9.22

1	2

1．田园气息的藤椅和金属材质的对比，形成了传统元素与现代时尚的空间对话。
2．老厂房原来的柱子和钢结构都被保留了下来，但它们都已经穿上了时尚的“外衣”。

1. Cane chairs with rural taste in contrast to metal texture form the spatial dialogue between traditional elements and modern vogue.
2. The columns and steel structures of the old plant are reserved, but they are now dressed in coat of vogue.

1．红色的钢结构是设计的一大亮点。
2．厚重的红色木门带着浓郁的传统色彩。
3．古朴的原木色家具。
4．装饰颜色主要以黑色的铁架、透明的大幅玻璃、大红色的置物架、原木色的桌椅构成，既有现代主义应有的热情，又不乏古典主义专有的理性。

1. The red steel structure member is a highlight of the design.
2. The red bulky wooden door is full of strong traditional taste.
3. Primitively simple wood colored furniture.
4. Black iron racks, transparent glass in large pieces, red shelves and original wood-colored chairs, which have the passion of modernism and rationality of classicism.

VEP Design
唯品设计

位置：上海市莫干山路50号2号楼CQL设计中心　电话：+86－021－62661555－818

唯品设计的设计师结合LOFT空间设计，采用玻璃幕墙、铝扣板、木材、透明大玻璃等多种现代材料，穿插运用，利用材料虚实、轻重、明暗等的对比手法，凝重中见轻快，使唯品设计本身在城市办公楼群体中形象鲜明突出。

整个办公楼共有两个楼面，一楼是唯品设计的大厅和独立办公室。设计师利用灰色的地面、透明玻璃、白色的墙打造了一个极富现代感的工作室。大厅内穿插着几块白色的板块，横竖交错的木板形成了一个虚空间，使空间的层次更加丰富。独立工作室的设计相当简洁，大幅的磨砂玻璃墙打破了传统建筑的封闭感，采光通透。

设计师非常巧妙地将通往二楼的楼梯设置在了一楼办公区外面的露天场所，全透明玻璃打造封闭式楼梯，这样的设计处理形成了内外视觉上的开敞状态、空间上的相互渗透和连续性，也达到了与整体空间一体化的效果。

二楼是一个开放式的办公区，设计手法与一楼截然不同。设计师充分利用二楼近十米的挑空，在工作室的一侧搭建了一个腾空的办公空间，将空间向纵向发展，改变了城市建筑群单调的水平线。同时为了令工作室更加宽敞明亮，设计师在屋顶的两侧设计了侧窗，敞亮的天棚设计代替了严实的房顶，自然采光代替了人造灯光，充分展示现代办公楼的时代特色。

唯品设计最独特之处在于二楼屋顶的独立露台，深褐色的木质地面铺材，与白色的墙面形成强烈的反差，显得亲切而随和。木制的矩形镂空空间与露台高低错落，内外空间渗透交融。

The designer of VEP DESIGN adopts glass curtain walls, aluminum buckled sheets, timber, transparent glass in large piece and other modern materials, which are used in alternation between lightness and heaviness, between light and dark and between false and true. These make VEP DESIGN itself outstanding among office building designs in the city.

The entire office building has two parts. The first floor is made of a hall and independent offices, where the designer uses grayish floor, transparent glass and white walls to create a studio full of modern taste. Several white boards interlaced within the hall, therefore a false space among the interlaced wooden boards. The design of independent studio is rather simple, and abrasive glass walls in large blocks break the sense of enclosure of traditional architecture and have good lighting.

The designer smartly puts the staircase leading to the second floor at an exposed place out of the office area of the first floor. The enclosed staircase made of transparent glass leads to the visual opening on both exterior and interior, mutual pervasion between spaces and continuity, and reaches the effect of unification with the entire space.

At the second floor is an open office, whose design is strikingly different from that of the first floor. The designer fully uses rising space of about 10 meters high to create a hanging office space at a side of the studio. Meanwhile, the designer designs side windows at two sides of the roof for the sake of acquiring brightness for the studio. The light and spacious ceiling takes the place of usual solid one and natural lighting takes the place of artificial lights, which fully shows the characteristic of office building of the time.

The most special point of VEP DESIGN lies on its independent gazebo on the roof of second floor. Its dark brown wooden flooring is strongly in contrast with the white walls, which seems friendly and easy. Rectangular space enclosed in holed wooden material is in alternate heights with the gazebo.

1. 设计师利用灰色的地面、透明玻璃、白色的墙打造了一个极富现代感的工作室。

1. The designer uses grayish floor, transparent glass and white walls to create a studio full of modern taste.

1	2	3

1、2. 大厅内穿插着几块白色的板块，横竖交错的木板形成了一个虚空间，使空间的层次更加丰富。

3. 全透明玻璃打造的封闭式楼梯形成了内外视觉上的开敞状态、空间上的相互渗透和连续性。

1、2. Several white boards interlaced within the hall, therefore a false space among the interlaced wooden boards.

3. The enclosed staircase made of transparent glass leads to the visual opening on both exterior and interior, and mutual pervasion between spaces and continuity.

1．设计师将空间向纵向发展，改变了城市建筑群单调的水平线。
2．向上延伸的楼梯。
3．大幅的落地玻璃窗。
4．设计师充分利用二楼近十米的挑空，在工作室的一侧搭建了一个腾空的办公空间。

1. The designer develops the space in depth, and alters the horizontal of the skyline of the city.
2. Staircase prolonging upward.
3. Grounded glass window in large pieces.
4. The designer fully uses rising space of about 10 meters high to create a hanging office space at a side of the studio.

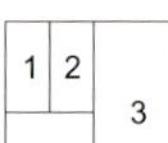

1．设计师在屋顶的二侧设计了侧窗，敞亮的天棚设计代替了严实的房顶，自然采光代替了人造灯光。
2．二楼是一个开放式的办公区，充分展示现代办公楼的时代特色。
3．独立工作室的设计相当简洁，大幅的磨砂玻璃墙打破了传统建筑的封闭感，采光通透。

1. The open and spacious ceiling takes the place of usual solid one and natural lighting takes the place of artificial lights.
2. At the second floor is an open office, which fully exhibits the features of modern office building.
3. The design of independent studio is rather simple, and abrasive glass walls in large blocks break the sense of enclosure of traditional architecture and have good lighting.

1	2

1．深褐色的木质地面铺材，与白色的墙面形成强烈的反差，显得亲切而随和。
2．木制的矩形镂空空间与露台高低错落，内外空间渗透交融。

1. Dark brown wooden flooring in strong contrast to white wall seems affable and easy.
2. Rectangular space enclosed in holed wooden material is in alternate heights with the gazebo.

Enlight Media
光线传媒

位置：上海市长宁路350号5楼　电话：+86－021－52378808

光线传媒的设计风格与其他办公楼的风格迥然不同，是以公司的文化理念为设计的基础，强调效率和人文环境的平衡，强调紧张与放松的统一。因而形成了多样性和个性化、倡导新时尚的办公空间。整体设计前卫、大胆、张扬，强调了视觉的冲击感。

从踏进工作室的那一刻起，五彩缤纷的色彩就牢牢地抓住了眼球。整个墙体通过几十种张扬的色彩和银灰色的金属被固定在一起，跃动的色彩在银色的基调中闪耀，纵向的延伸感更是特别强调了空间的高度，令空间充满着律动的节奏感。如此多的色彩使得该建筑的每个部分都显得独一无二、与众不同，绝对不会给人造成无聊的视觉印象。整体的建筑风格处理简洁，富有时代的建筑空间特征，设计方法具有条理，节奏对比强烈：挺拔宏大的楼身和宽敞通透的大厅和回廊融和得天衣无缝；白色的休闲椅与黑色的钢琴漆搭配得相得益彰；充满现代感的大幅玻璃和朴素古朴的木制家具对比强烈。

设计师更是独具匠心地将造型、功能与结构协调统一在设计中，充分考虑造型与功能的有机结合。从使用功能和结构要求考虑，将整个工作室划分为不同的区域，办公区、会议室、休闲区、接待区等。在楼梯的造型上也形成了视觉中心，时而曲折向上，时而蜿蜒向前，时而又盘旋而上。整个设计极具现代感和时尚感。

The design style of ENLIGHT MEDIA is strikingly different from other office buildings. Its style is based on its enterprisal culture and ideas, which emphasize the equilibrium between working efficiency and human environment, and emphasize the union of tension and relaxation. Therefore, diversity and individuality are formed to create novel and voguish office space. The entire design is van guarded, bold, and extensive and emphasizes visual impact.

Multi-colors attract people's eyes immediately after stepping into the studio. The whole walls are fixed together by dozens of colors and silver-grayish metals. Springing colors sparkle within the key color as silver, and horizontal extension emphasizes the height of the space, which makes the space full of pulsatile rhythm. So many colors make every part of the architecture seems unique. The style of the entire architecture is simple, full of architectural space character of the time, and strong in rhythm and contrast. Grandiose body and spacious hall and cloister are perfectly melted; white leisure chairs and black piano are well matched; and modern glass in large pieces is strongly in contrast with primitively simple wooden furniture.

The designer artfully puts the forms, structures and functions into the design and fully considers the organic union of formation and function. Based on this, the entire studio is divided into different areas: office, meeting, leisure and reception, etc. A visually focal point is formed with respect to the formation of staircase, which is partially zigzagged upward, partially winding forward and partially scroll upward,. The entire design is full of modernism and vogue.

P196-P197

1

1．整个墙体通过几十种张扬的色彩和银灰色的金属被固定在一起。

1. The whole walls are fixed together by dozens of colors and silver-grayish metals.

1 | 2

1．从踏进工作室的那一刻起，五彩缤纷的色彩就牢牢地抓住了眼球。
2．高挑的空间为整个工作室的设计提供了得天独厚的优势。

1. Multi-colors attract people's eyes immediately after stepping into the studio.
2. High rising space offers superiority to the entire studio.

ENLIGHT MEDIA

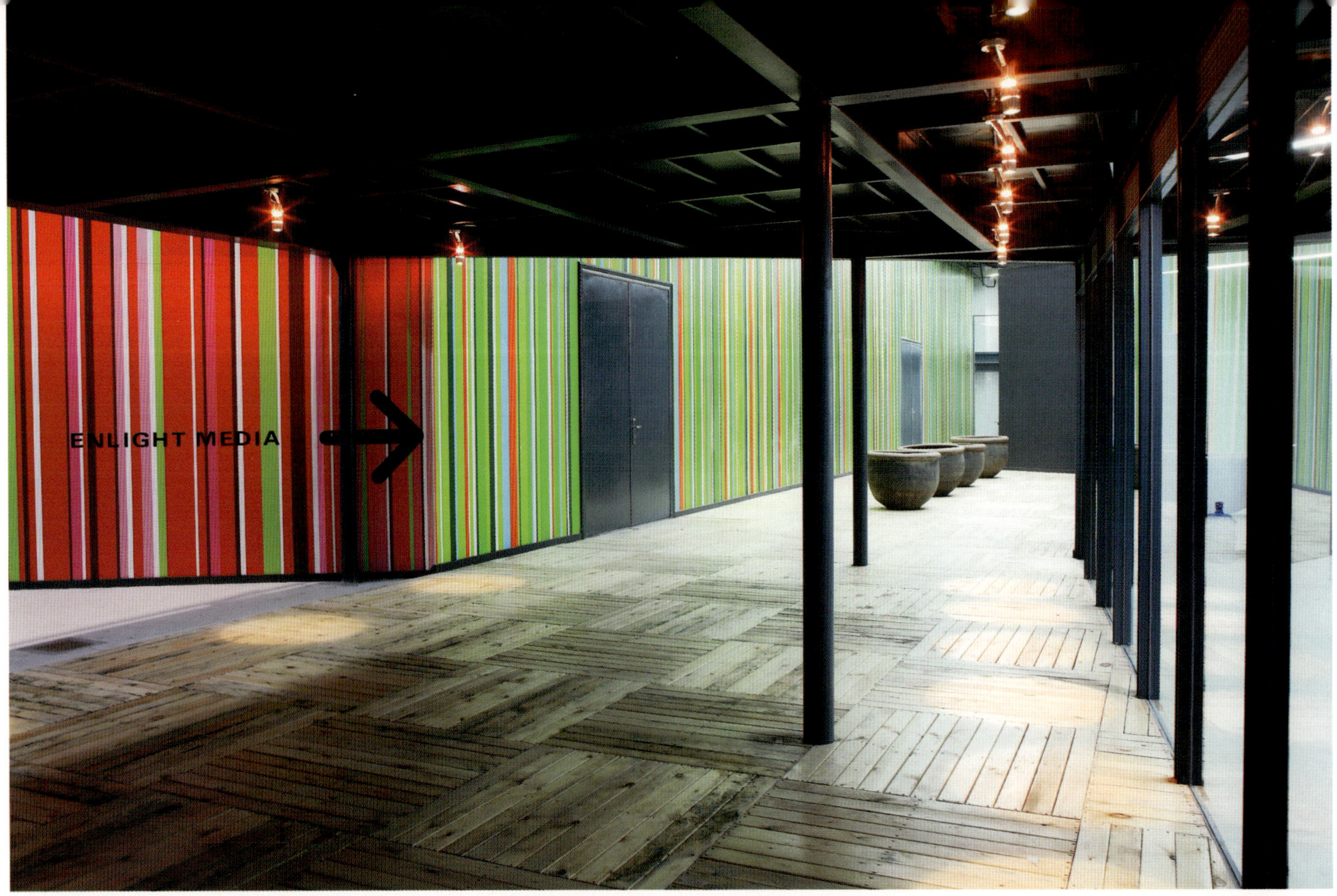
ENLIGHT MEDIA

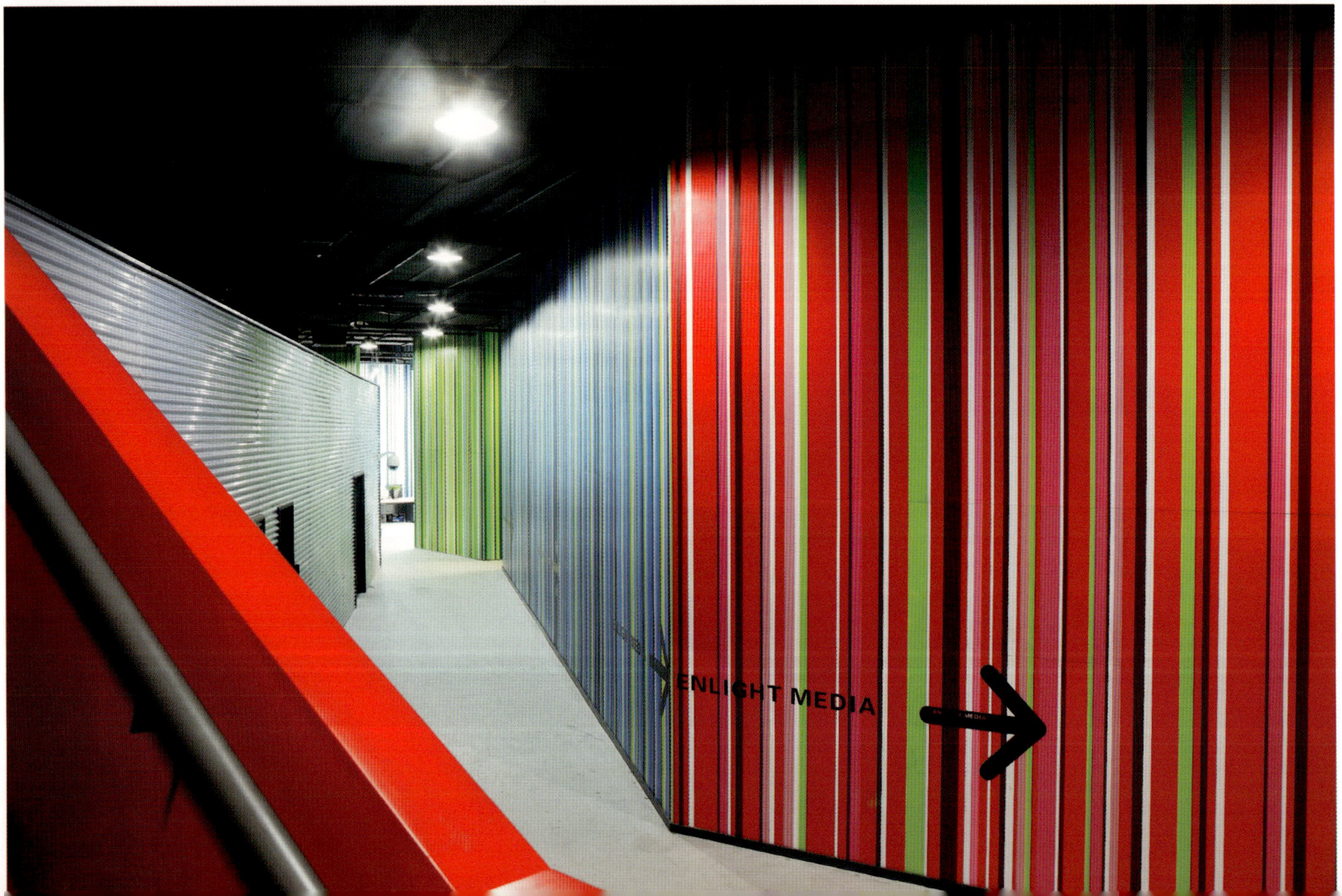
ENLIGHT MEDIA

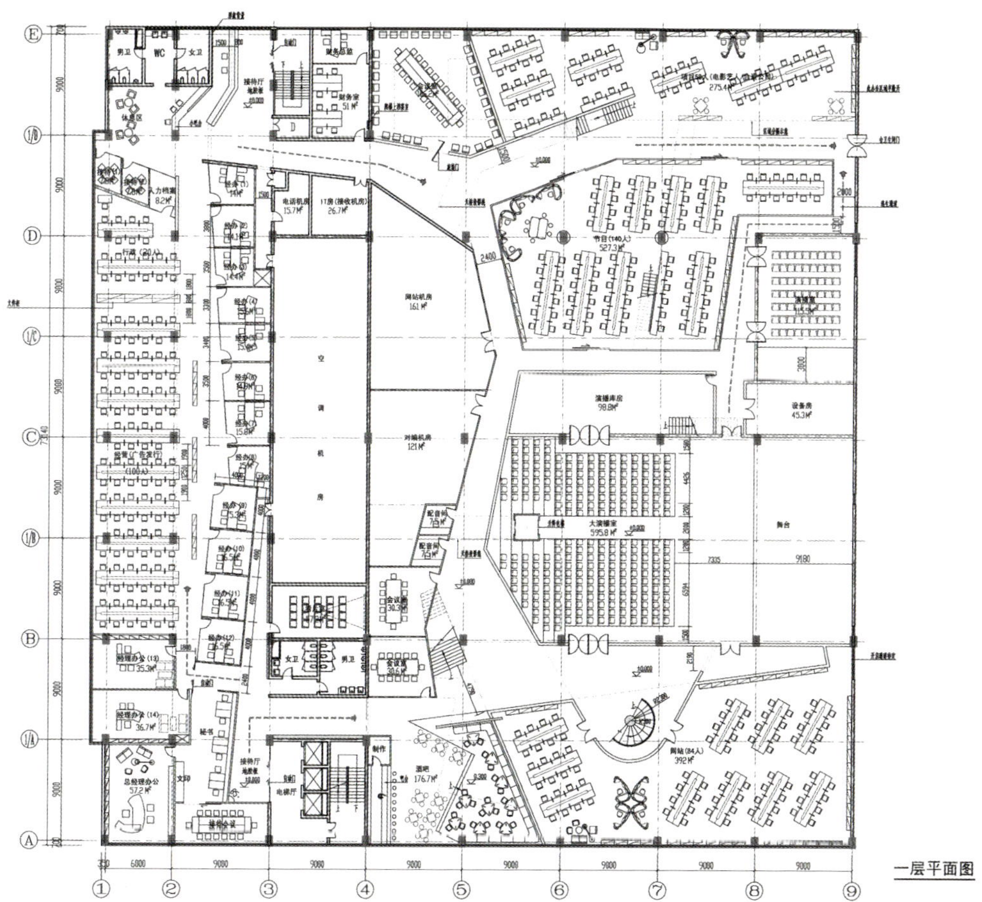

1	
2	3

1. 纵向的延伸感更是特别强调了空间的高度，令空间充满着律动的节奏感。
2. 跃动的色彩在银色的基调中闪耀。
3. 夸张的箭头指示强调了个性化的设计。

1. Horizontal extension emphasizes the height of the space, which makes the space full of pulsatile rhythm.
2. Springing colors sparkle within the key color silver.
3. Exaggerate arrow signs emphasize the individualistic design.

1. 加长的镜面反射着五彩缤纷的墙面，令空间更宽敞。
2. 曲折向上的楼梯。
3. 设计师更是独具匠心地将造型、功能与结构协调统一在设计中，充分考虑造型与功能的有机结合。

1. Lengthened mirror reflects colorful wall surface, which makes the space seem more spacious.
2. Staircase zigzagging upward.
3. The designer artfully puts the forms, structures and functions into the design and fully considers the organic union of formation and function.

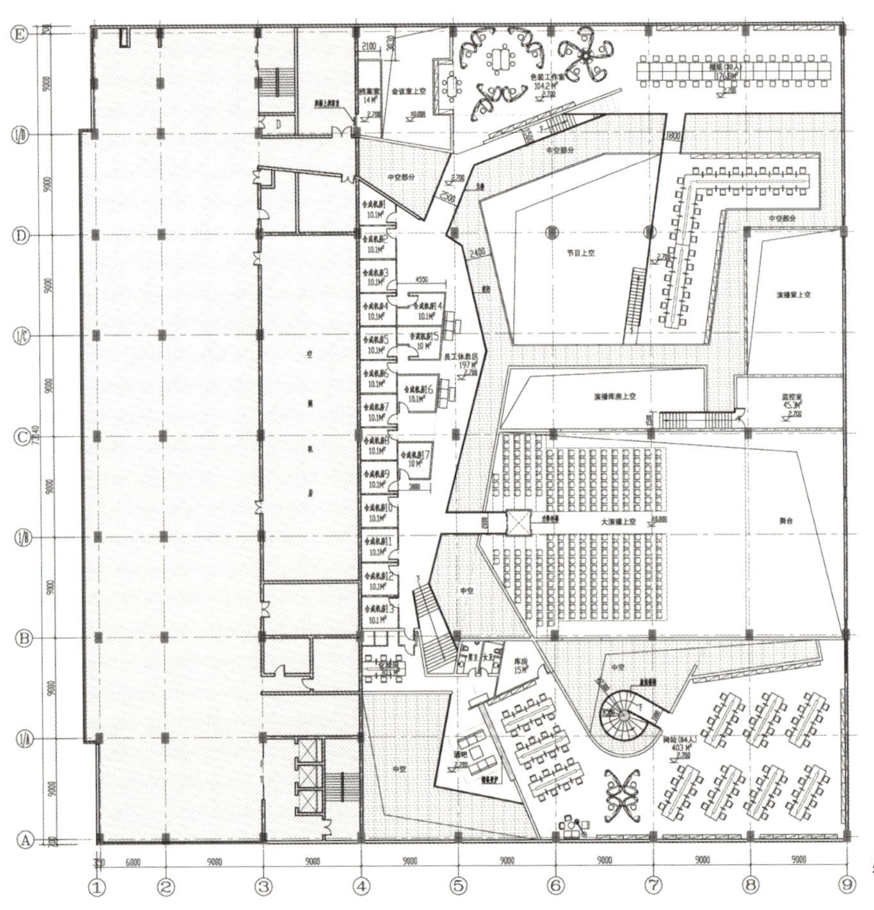

局部二层平面图

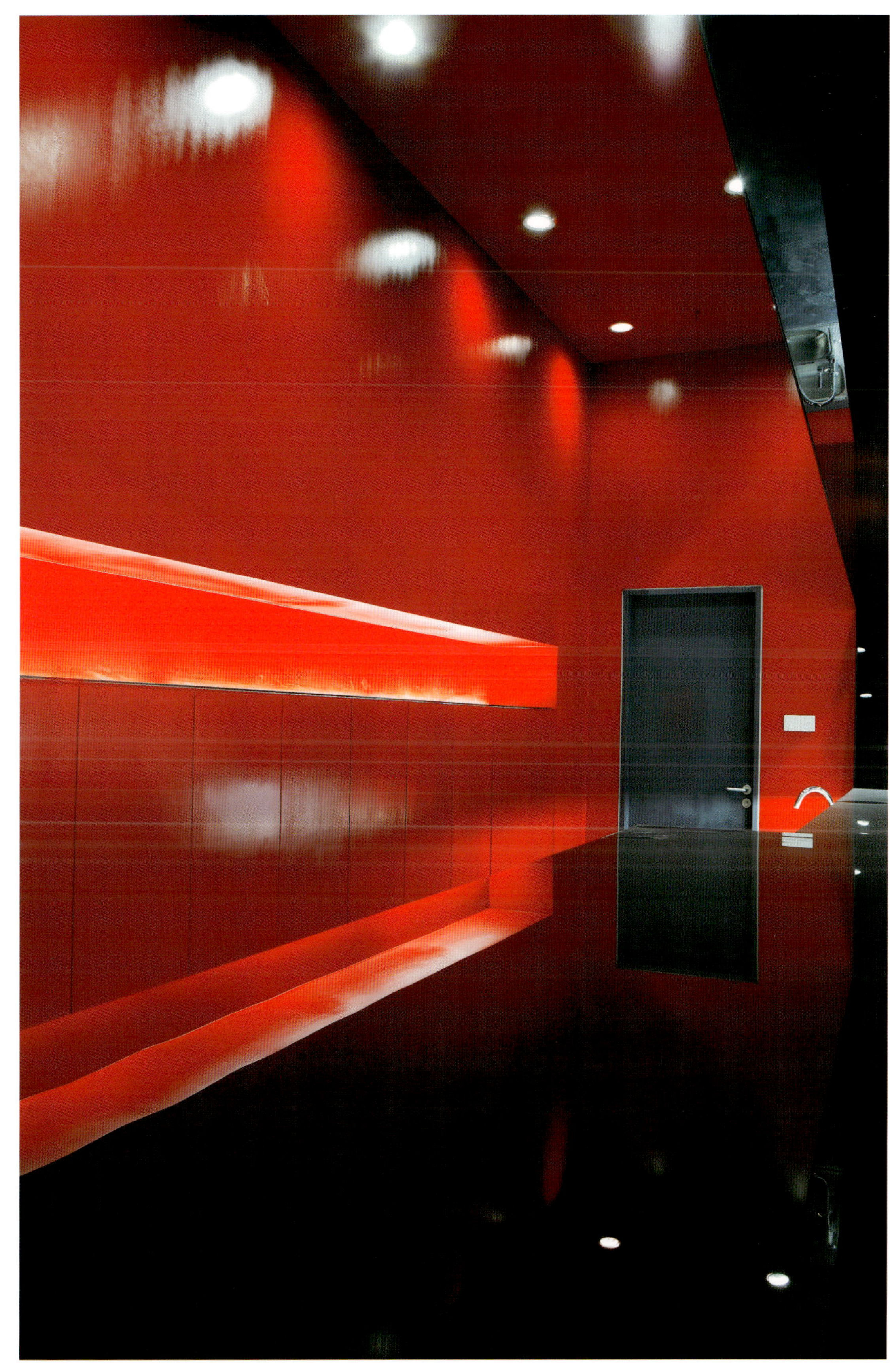

1	3
2	

1．设计师特别强调了竖向线条。
2．回字形的会议室同样充满了色彩的韵律。
3．火红色的墙体显得热情、张扬。

1. The designer particularly emphasizes on vertical lines.
2. Meeting room in double squares with a same center is full of colorful rhythm.
3. Flamboyant reddish wall blocks seem passionate and exaggerate.

P206-P207

1	2	5
3	4	

1、2、3、4.
时而曲折向上，时而蜿蜒向前的楼梯和走廊。
5. 白色的休闲椅与黑色的钢琴漆搭配得相得益彰。

1、2、3、4.
Staircase partially zigzagging upward and partially winding forward.
5. White leisure chairs and black piano are well matched.

1	3
2	

1. 倾斜的银灰色的金属墙面打破了单调感，颇具现代时尚感。
2. 楼道一角。
3. 盘旋而上的楼梯。

1. Inclined silver metal wall face breaks monotony and is full of modernism.
2. A part of the staircase.
3. Staircase twisting upward.

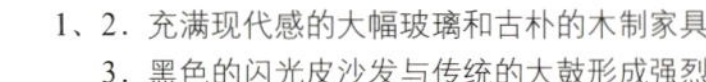

1、2. 充满现代感的大幅玻璃和古朴的木制家具对比强烈。
3. 黑色的闪光皮沙发与传统的大鼓形成强烈的对比。

1、2. Modern glass in large pieces is strongly in contrast with primitively simple wooden furniture.
3. Black flashing leather sofa is strongly in contrast with traditional drum.

P212-P213

1. 白色的休闲椅搭配玻璃的晶莹剔透更显得简洁明朗。
2. 如此多的色彩使得该建筑的每个部分都显得独一无二、与众不同。绝对不会给人造成无聊的视觉印象。

1. Combination of white leisure chairs and translucent glass seems cleanly and lucid.
2. So many colors make every part of the architecture seem unique and never let people feel visually senseless.

Deng Kunyan Architectural Design

登琨艳建筑事务所

位置：上海市杨树浦路2218号　电话：+86－021－65683498

登琨艳的建筑事务所位于他所规划的滨江创意产业园（也称作大杨浦）的最外端，那里原是1921年所兴建的上海电站辅机场，当初是美国GE电子在亚洲投资最大的电子工厂，无论从历史还是建筑角度，那里都是极具文化价值的工业遗址。

事务所的占地像是个拉长的L字的造型，仅有两层的建筑保留了完好的尖顶木架梁体，事务所的入口设置在了L字聚拢的夹角处，恰好正对南面。登琨艳将入口设计得犹如一个用于展示的阳光棚，地面上的透光地板错落布局，在夜间辉映出神奇的灯光效果，上层则是一个露台，连着他自己的办公室，与周围的绿荫相连。

往里就是最具特色的玄关区域，登琨艳将那里设计得犹如一个舞台，空间被整个拉空，白色主导了那里的颜色，墙上的圆洞、刻意砸开的一圈天花板、漆成蓝色的弧形身的大水缸，都赋予了空间更多的线条。斜置的镜子占据了玄关通向二楼办公区域楼梯一侧的整面墙，镜子幻化出的空间与真实的存在形成一个虚实交互的界面，让人产生奇妙的感觉。天花处垂挂下大小错落的球灯，令整个玄关区域犹如梦境一般。

自玄关上到建筑上层，分左右两翼展开，一侧是登琨艳自己的办公室，另一侧是事务所其他设计师的办公区域，连接两个区域的是一个过道平台。平台右侧是事务所其他设计师的办公区域，恢弘的空间裸露着建构的梁体，人在其中显得渺小，而如此气势又是人为而成，处在这样的氛围中办公，必能激发起更多设计创造的智慧。

在空间过渡的区域，原有的建筑被很好地保留形成一个界面，底下是员工休憩的区域，上方是一个阁楼。从搭建的阁楼上，有一个非常棒的眺望区域。在接近建筑梁体的部分，更设置了榻榻米的休憩场所，方便设计人员进行休息。

过道左侧是通向登琨艳独立办公区域的，入口用了他一向喜欢的金属丝网材料，登琨艳赋予了原本硬质的材料以轻薄的视觉效果，与清冷的大理石台面形成了对比。推开巨大的木移门，入眼的是办公室如画般的休息区域，窗外的绿意透过落地窗台漏了进来，于宏大中透出静谧。登琨艳的办公桌颀长无比，选用了整面的大理石台面，气势宏大，其实也非常符合大项目图纸的铺排。

Deng Kunyan Architectural Design is located on the edge of Binjiang Originality Industry Garden which is programmed by him, a place where there used to be a airport in 1921. At that time, it is the plant of electric products which was funded and established by American General Electrics. No matter in historical perspective, or in architectural perspective, the place is industrial site full of cultural value. The area is much like a strengthened "L", on which a building with only two stories has well reserved pointed roof and wooden structured beams and girders. The entrance of the office is set at the inner angle of the "L" and faces the south. Deng Kunyan designs the entrance as a pergola (awning) for display. The light reflecting flooring is laid in crosswise pattern, on which fantastic lighting reflects in the evening. Above it there is a gazebo which is connected to his own office adjacent to surrounding green shade.

Going deeper inside, one can find the zone full of specialty, through Deng Kunyan's design, it looks like a stage, the space is drawn empty entirely and white dominates the colors, the wholes on the wall, deliberately opened ceiling and flooring in circles and arced aquarium all infuse the space with more lines and strips. Slanting mirror occupies a whole piece of wall near the staircase where the lobby leading to the office area of the second floor. An interface is formed between the space mirrored in the mirror and the true space, which makes people feel fantastic. Ball lamps in different size fall from the ceiling, which makes the whole lobby as if in a dreamland.

From the lobby to its upper level, the space is unfolded into left and right, on one side is Deng Kunyan's own office, and on the other side is the office area for other designers, and a flat corridor links the two areas. There are many exposed beams in the vast space, within which people seem tiny. It is human beings who created such vast space and atmosphere in which people can be inspired much to contribute design wisdom.

In the transitional area, the original architecture is perfectly preserved to form an interface, beneath it is a resting area for the employees, and the above is an attic in which there is a fine area for overlook. At the part adjacent to the beam members, resting place on tatami is set to accommodate the employees.

Material as metal wires are used to the entrance to Deng Kunyan's own office, and Deng Kunyan infuses solid hard material with visually light and thin effect, which is in contrast with plain and cold marble desktop. After opening the wooden sliding door, picturesque resting area of the office falls into eyes. The green out of window leaks from grounded window sill, and adds tranquility to the magnificence. Deng Kunyan's desk is incomparably long, and its top is made of entire piece of marble, which is suitable for the spread of drawings and blueprints of great project.

1

1. 原本 GE 的工厂有着非常优秀的园林环境，登琨艳借此添加一些建筑原材，并进行了构成布局，一个个美丽的景观就这样完成了。

2. The former American General Electrics' plant was in excellent landscape, and Mr. Deng adds some architectural materials to it to forms a more beautiful landscape.

1	2	3
		4

1．巨大的推门是登琨艳自己办公室的入口，显得气势宏大。
2．恢弘的空间裸露着建构的梁体，人在其中显得渺小，而如此气势又是人为而成，处在这样的氛围中办公，必能激发起更多设计创造的智慧。
3．登琨艳办公区域的入口用了他一向喜欢的金属丝网材料，登琨艳赋予了原本硬质的材料以轻薄的视觉效果，与清冷的大理石台面形成了对比。
4．透过圆洞，你可以清晰地看见玄关区域的动静，透过半敞的木门，你可以通晓办公区域的情境，空间的气韵就这样流动了起来。

1. The huge door links the entrance to Mr. Deng's own office room, which seems grandiose.
2. There are many exposed beams in the vast space, within which people seem tiny. It is human beings who created such vast space and atmosphere in which people can be inspired much to contribute design wisdom.
3. Material as metal wires are used to the entrance to Deng Kunyan's own office, and Deng Kunyan infuses solid hard material with visually light and thin effect, which is in contrast with plain and cold marble desktop.
4. Through the round hole, one can clearly view the movement within the lobby, and through the half open door, one can learn the conditions in the office area.

1. 镜子与楼梯形成一个虚实交互的界面。
2. 登琨艳将玄关设计得犹如一个舞台，空间被整个拉空，白色主导了那里的颜色，墙上的圆洞、刻意砸开的一圈天花板、漆成蓝色的弧形身的大水缸，都赋予了空间更多的线条。
3. 斜置的镜子占据了玄关通向二楼办公区域楼梯一侧的整面墙，镜子幻化出的空间与真实的存在让人产生奇妙的感觉。

1. An interface is formed between the mirror and the staircase.
2. Through Deng Kunyan's design, the lobby looks like a stage, the space is drawn empty entirely and white dominates the colors, the wholes on the wall, deliberately opened ceiling and flooring in circles and arced aquarium all infuse the space with more lines and strips.
3. Slanting mirror occupies a whole piece of wall near the staircase where the lobby leading to the office area of the second floor. An interface is formed between the space mirrored in the mirror and the true space, which makes people feel fantastic.

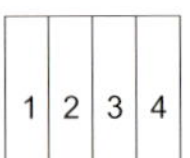

1. 通向登琨艳办公区域的过道。
2. 接近建筑梁体的一个阁楼平台，设置了榻榻米的休憩场所，方便设计人员进行休息。
3. 原有的建筑被很好地保留，搭建的阁楼形成一个非常棒的眺望区域。
4. 从上部俯瞰，天花处垂下的球灯令整个玄关区域犹如梦境一般。

1. hallway leading to Mr. Deng's office room.
2. At the part adjacent to the beam members, resting place with tatami is set to accommodate the employees.
3. the original buildings are well preserved, and the built attic forms a good area for overlook.
4. Ball lamps in different size fall from the ceiling, which makes the whole lobby as if in a dreamland.

1．过道通往事务所设计师的办公区域。
2．青砖墙面的处理非常出色，随机却不随意的排列显出韵律感。
3．自玄关上到建筑上层，分左右两翼展开，一侧是登琨艳自己的办公室，另一侧是事务所其他设计师的办公区域，连接两个区域的是一个过道平台。

1. hallway leading to other designers' office area.
2. Blue bricked wall is well treated, and the random array reveals rhythm.
3. From the lobby to its upper level, the space is unfolded into left and right, on one side is Deng Kunyan's own office, and on the other side is the office area for other designers, and a flat terrace links the two areas.

P224-P225

1	2
	3

1. 从阁楼平台俯瞰整个玄关区域。
2、3. 登琨艳的办公桌颀长无比，选用了整面的大理石台面，气势宏大，其实也非常符合大项目图纸的铺排。

1. Bird's eye view of the entire lobby area from the attic.
2、3. Deng Kunyan's desk is incomparably long, and its top is made of entire piece of marble, which is suitable for the spread of drawings and blueprints of great project.

1	2	3

1. 办公室内的休息区域犹如画幅一般，宏大中透着静谧。
2. 通向上层阁楼平台的楼梯，其构成像是一个木质的装置艺术作品。
3. 从阁楼平台俯瞰设计师们的办公区域。

1. Picturesque resting area of the office is both grand and tranquil.
2. the staircase leading to the attic on upper floor is just like a wooden decoration in form.
3. Bird's eye view of the other designers' office area.

Horizon Design
群裕设计

位置：上海市杨树浦路2218－12号 电话：＋86－021－58872932

群裕设计（潘冀联合建筑师事务所）在台湾已有25年历史，它在上海的办公地点位于登琨艳规划的大杨浦内。

整个旧厂房空间的再利用遵循了这样的理念，尊重既有空间与历史文脉，将工厂的外墙（包含开窗）、屋顶、钢结构全部保留，并施以机能上的翻修（例如漏水）及清洁。因此所呈现的空间，让人一进入就能感受到旧有的建筑与新的室内设计两者间的相互结合。

既有的建筑外壳，其气密性的不佳导致了空间隔音隔热效果的不良，设计方案基于节能及舒适性等因素的考量，将整个厂房视作一个“半开放半户外”的空间，保留它流畅自然的通风与采光，而办公空间、会议室、图书室等所有办公所需的空间，则以“建筑中的建筑”的型态交错出现，最终形成了我们所见的两个“盒子”—方形玻璃屋用作设计工作区，椭圆形“盒子”则是脑力激荡及会议区，一方一圆，一透明清澈一青砖敷体，形成对照。

空间的多样可能也丰富了建筑的弹性。会议区的活动隔扇拆卸后，便可与图书区连成一气，图书区的架高平台摇身一变成为表演舞台，会议区则成了观众席。图书区被定义成解压、交谊、能源的补充区，在那里使用者可以进行知识能源（图书）、体力能源（食物、茶水）的补给，此区域恰好有一个楼梯通往楼上，整个区域顺势垫高，垫高的构件又成了办公室的储藏室。整个办公空间还留置了多功能的夹层区，作为休憩、讨论的空间，同时也是未来办公空间发展的扩充空间。方形玻璃屋的楼上还预留了一些空间以便将来扩充的需要。

Horizon Design (Pan Yi Associated Architects office) has a history of 25 years in Taiwan, and its quarter in Shanghai is located in the great Yangpu region planned by Deng Kunyan.

The reutilization of the entire old plant follows such deaign idea respecting both the existing space and historical culture to reserve all the exterior walls(including windows), roofs and steel structures and to repair and clean organically. So the present space can give people who immediately within it some sense about the combination of the old architecture and new interior design.

The existing architectural exterior's low air compactness leads to the inferiority of sound and heat insulation. Based on consideration on factors as energy saving and comfort ability, the design treats the entire plant as a semi-opening and semi-outdoor space, and reserves its natural fluent ventilation and lighting. While the office, meeting room, library and other required spaces appear alternatively and finally form what we see as two “boxes”, rectangular glass room is used as an area for designing work and the other ellipse “box” is used as brain motivating and meeting area. There is a sharp contrast between a squared form and round form as well as between transparent glass and black bricks.

The diverse spaces possibly enrich the springiness of architecture. If the detachable door of the meeting room is detached, a unit space is formed since the meeting room is in connection with the library. Then the elevated platform in the library immediately becomes a performing stage and the meeting area becomes an auditorium. Library is defined as an area for relaxation, exchange and energy supplement mentality (books) and vitality (food, tea and etc.) are supplied. In this area, a staircase leading upstairs, and under the staircase there is storehouse. The entire office space keeps multi-function interlaying zones for resting and discussion as well as future expansion of the office space. The upper floor of the rectangular glass room preserves some spaces for future expansion too.

1

1．椭圆形会议区的外墙使用了从上海本地老房子拆除时所废弃的青砖。后侧的图书区所用的枕木也是老建材，整个抬高的体量其实也是办公室的储藏室。

1. The exterior wall of oval meeting area uses the blue bricks detached from the local old houses. The crossties used in library area are also old building materials. The whole elevated part is actually storeroom of the office.

1. 玻璃方盒的办公区域之上设置了一个个由耐火砖构筑的模型台，展示公司各个出色的项目。
2. 乳清色的地面使用环氧树脂建造，考虑到了不易起尘的环保效果，保证了室内的空气质量，而且也比较容易清洁。
3. 基于建筑本身不良的气密性，设计者采用了“半开放半户外”的空间构成概念，在保证了现代办公需要的前提下，又将老建筑的特色得以保留，两全其美。

1. Model desk made of firebricks is placed on the glass boxed office area, and it is used to display excellent projects practiced by the company.
2. Milky floor is made of colophony, considering that it won't easily cause dust, is sure to guarantee the quality of air interiorly and is easily cleaned.
3. Considering the factors that the building is low in air compact, the designer adopts semi-open & semi-outdoor space concept, which both assures the requirement of modern office and preserves the features of the old building.

1	2	3

1. 老厂房的框架并没有作过多的修缮，只是进行了一些清洗，现代与历史文脉相互串联了起来。
2. 空间外墙的侧面本是一条长长的过道，设计者将那里规划成了洗手间和淋浴房，方便员工使用。
3. 图书区域凭借一铸铁楼梯与方形玻璃的办公区域相连。

1. The structure and frameworks are repaired less, only some cleaning is done. Hence the historical containing and culture are preserved.
2. At the side of the exterior wall of the space is a long hallway, and the designer set a lavatory and shower room which are for the employees' convenience.
3. The library area is linked to squared office area enclosed in glass through a cast iron staircase.

P234-P235

1	3
2	

1、2. 此区域为解压、交谊、能源的补充区，它连通方形玻璃盒子的办公区域，也通过变成舞台的形式与椭圆形的会议区形成交互。

3. 办公区域内的隔断刻意地使用了现出锈色的金属板，外层再添加防锈处理，与玻璃盒子的外墙形成鲜明对比。

1、2. It is defined as an area for relaxation, exchange and energy supplement, and it is linked to glass boxed office area.

3. The partition within the office area deliberately uses rusted metal board with rust-proof coating, which is in contrast with glass box exterior wall.

Horizon

ILX MANDARIN
艾尔克斯传媒公司

位置：上海市南苏州路1305号1楼　电话：+86－021－63593631

艾尔克斯咨询有限公司（ILX MANDARIN）是一家从事媒体、医药的咨询公司。它地处南苏州河边的老仓库，显得非常有创造力。

公司的设计由A00完成。整个空间使用了大量的木质材料，温润而凝重，与所保留的老仓库的建筑细节相得益彰。从玄关走入空间，最先入眼的是透明玻璃隔断中间装设的公司的LOGO，那即是一种导引，也让来人不易碰撞到。一侧的区域，四把沙发即构成会客区域的简洁布局，其实主要是源于空间中原有老建筑的丰富细节，如此规划才让人感觉非常妥帖。

临近会客区的是公司高层人员所使用的独立办公区域，设计师架起方格铸铁的隔断，配上磨砂的白色玻璃，既保证了老建筑空间的采光需要，又保留了空间的私密性，两全其美。独立的办公区域两侧分别设计了木质的搁柜，中间则使用了玻璃。设计师特意使用了玻璃，而不是镜面，同时又在相邻的办公区域中进行了镜像的格局设置，将一个真实的空间幻化成虚拟的空间，犹如进行了一场空间的魔术。

再往里就是主办公区，从开敞的门向内看，横平竖直的梁体柱体使得空间非常有气势。整个区域内的办公桌与木质搁柜使用了统一的模块化设计，整齐划一，其悬空搁置的方式也显得非常新奇，而颀长的桌面则非常适合工作的需要。

ILX MANDARIN is a consultation company which deals with media and medicine. It is located near south SuZhouhe River in an old warehouse, which is extremely creative.

The design is implemented by A00. The entire space consumes large quantity of woody material which seems soft and dignified, and this is compatible with the preservation of the old warehouse. Going into the space through the lobby, the first to be falling into eyes is the LOGO between transparent glass partitions; it is a sort of guide. On one side is an area in which four pieces of sofa constitute a simple guest receiving area. It is because of deriving from the rich old architectural details that makes the design very appropriate for people.

Near the guest receiving area is the independent office area for top administrative personnel, checkered cast-iron partition in companion with abrasive white glass both assures the need of lighting from the old architecture and preserves the privacy of the space. Wooden cabinets are set on both sides of the independent office area, and glass is used in the middle of the area. The designer deliberately uses glass instead of mirror, meanwhile he designs mirror reflection at the adjacent office area and turns a real space into an illusive one.

If going further, one will see the main office area. On the interior, horizontally straight beams and vertically straight pilasters seem very vigorous. The desks and wooden cabinets within the entire area are uniformly modularized in design. The hanging settlement of them is very novel and peculiar, and the long desktop is quite suitable for the work.

P236-P237

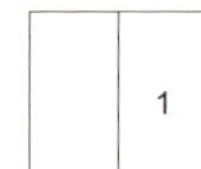

1．四把沙发即构成会客区域的简洁布局，其实主要是源于空间中原有老建筑的丰富细节，如此规划才让人感觉非常妥帖。

1. Four pieces of sofa constitute a simple guest receiving area. It is because of deriving from the rich old architectural details that makes the design very appropriate for people.

P238-P239

1	2	3
		4

1．公司高层人员有着独立的办公区域，设计师架起方格铸铁的隔断，配上磨砂的白色玻璃，既保证了老建筑空间的采光需要，又保留了空间的私密性，两全其美。
2．从洗手间透过玻璃窗看会客区域。
3、4．独立的办公区域两侧分别设计了木质的搁柜，中间则使用了玻璃。设计师特意使用了玻璃，而不是镜面，同时又在相邻的办公区域中进行了镜像的格局设置，将一个真实的空间幻化成虚拟的空间，犹如进行了一场空间的魔术。

1. Near the guest receiving area is the independent office area for top administrative personnel, checkered cast-iron partition in companion with abrasive white glass both assures the need of lighting from the old architecture and preserves the privacy of the space.
2. a view of guest receiving area from through lavatory window.
3、4. Wooden cabinets are set on both sides of the independent office area, and glass is used in the middle of the area. The designer deliberately uses glass instead of mirror, meanwhile he designs mirror reflection at the adjacent office area and turns a real space into an illusive one.

1	3	4
2		

1．办公区域全景。
2．办公桌与木质搁柜使用了统一的模块化设计，整齐划一，其悬空搁置的方式也显得非常新奇。
3．横平竖直的梁体柱体使得空间非常有气势。
4．办公桌侧影，颀长的桌面非常适合工作的需要。

1. overall view of office area.
2. The desks and wooden cabinets within the entire area are uniformly modularized in design. The hanging settlement of them is very novel and peculiar.
3. Horizontally straight beams and vertically straight pilasters seem very vigorous.
4. The long desktop is quite suitable for the work.

A00 Architecture
A00建筑设计公司

位置：上海市南苏州路1305号1楼　电话：+86－021－63276213

A00是家很有活力的建筑事务所，合伙人都是来自加拿大的年轻建筑师。仅从公司的外立面就能清楚地看到大会议室的所有细节，事务所活跃的气氛由此可见一斑。

A00的设计师Raefer（潘朝阳）是个健谈的人，也是个非常有创造力的建筑师，他总是将“玩”与建筑空间设计联系在一起。A00的空间就是他所“玩”的一个成果，当初他们租下这间仓库，设计完成之后觉得自己使用显得有些大，因此与艾尔克斯（ILX MANDARIN）共同使用了这个空间，可谓物尽其用。

办公室的入口犹如一个时光隧道，长长的，幽静而深暗，让人仿佛穿越了时空。Raefer喜欢用原始的材料，木也就是成了空间的宠儿，从地面延续到墙面。结构复杂的建筑给了Raefer以灵感，他将平淡无奇的入口设计得上下错落，形成非常有趣的区域。楼梯自下而上，再自上而下地进入接待前台。前台的空间其实是一个通道，设计师保留了这部分结构，并架设起一个天花，形成透光的阳光平台，非常出彩。仔细地端详漆成暗红色的墙体上，浮雕着装饰性的楼梯，让人错乱了空间的构成。

A00的会议室非常有趣，像是一间演播厅，墙面上的装饰犹如是幅油画，你可以在桌面上涂涂画画，你的想法能在第一时间传达给你的聆听者，因为亚克力的桌面当初就是因为这个原因而设计的。

公司的洗手间也非常有趣，你不必费心寻找哪里是男用哪里是女用，因为压根洗手间就只有一个共用的空间，只是每个隔间都是独立的，使用上也没什么不便，给人的感觉很好“玩”。

A00 is an active architecture office, and its copartners are all young architects who come from Canada. From the exterior elevation of the building one can see the details of the grand meeting room as well as the active atmosphere within the office.

The designer from A00, Raefer (Pan Chaoyang) is a conversational person and a creative architect as well. He always associates “playing” with architectural space. A00 is just a fruit of his playing. At the beginning of its establishment, they rented this warehouse, and it seems bigger after designing, so its space is shared with ILX MANDARIN.

The office entrance is like a time tunnel which is long, tranquil and dark, and people will feel they are going through time when going through it. Raefer prefers original material, so wood becomes favorable for the space, and can be seen from on the floor to on the wall. Complicated structure inspires Raefer who puts ordinary entrance into interesting area. The staircase in the first leads upwards, then leads down and finally reaches the reception desk. The space for reception desk is actually a channel which is reserved by the designer and a ceiling is set to form a special light-penetrating platform. If one carefully looks at painted wall in dark red, he will see decorative staircase in relief, which confuses people about the composition of the space.

The meeting room of A00 is rather interesting, and it is like a show or performing hall and the decoration on the wall looks like a piece of painting. You can the desktop can be used to scrawl on or paint on and your idea can be conveyed to your listener at the first place. The polymer desktop is so equipped just because of this reason.

The lavatory of the company is interesting too you don't need to care about which closet is for ladies and which for gentlemen it has a shared space in which every closet is independent and can serves both male and female. The convenience bring forth by such design is enjoyed by people.

P242-P243

	1

1．木是空间的主角，从地面延续到墙面。

1. Wood becomes favorable for the space, and can be seen from on the floor to on the wall.

1	3
2	

1．A00的入口犹如一个时光隧道，长长的，幽静而深暗，让人仿佛穿越了时空。
2．公司的前台区域。
3．前台空间其实是一个通道，设计师保留了这部分结构，并架设起一个天花，形成透光的阳光平台，非常出彩。

1. The office entrance is like a time tunnel which is long, tranquil and dark, and people will feel they are going through time when going through it.
2. Reception desk area.
3. The space for reception desk is actually a channel which is reserved by the designer and a ceiling is set to form a special light-penetrating platform.

1	2

1．漆成暗红色的墙体上，浮雕着装饰性的楼梯，
让人错乱了空间的构成。
2．结构复杂的建筑给了A00的设计师以灵感，将
平淡无奇的入口设计得上下错落，形成非常有趣的
区域。

1. The wall painted in dark red and decorative staircase with relief can confuses people about the composition of the space.
2. The designer is inspired by the architecture complicated in structure so much that he puts the ordinary entrance area into a well proportioned one which becomes very interesting.

1．大会议室内景。
2、3．纵深的直线是空间的构成要素，它让空间在视觉上显得更加深远。
4．细横木条的百叶门有些东南亚的感觉，比例和尺度上却非常恢弘。

1. indoor scene of the grand meeting room.
2、3. Line in depth is the constituent element of the space, and it lets the space seem more deeper visually.
4. Thin blind door with horizontal battens bears Southeast Asian taste, and its proportion and scale are much magnified.

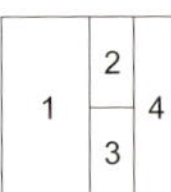

1、办公室一角。

2、3、A00的会议室非常有趣，像是一间演播厅，墙面上的装饰犹如是幅油画，你可以在桌面上涂涂画画，你的想法能在第一时间传达给你的聆听者，因为亚克力的桌面当初就是因为这个原因而设计的。

4、几何分隔的柜子成了空间中的一种隔断。

1. A glimpse of the office room.

2、3. The meeting room of A00 is rather interesting, and it is like a show or performing hall and the decoration on the wall looks like a piece of painting. You can the desktop can be used to scrawl on or paint on and your idea can be conveyed to your listener at the first place. The polymer desktop is so equipped just because of this.

4. Geometrically divided cabinet becomes a sort of partition of the space.

图书在版编目(CIP)数据

仓库办公室 / 小白摄影；钟音撰文.-沈阳：辽宁科学技术出版社，2007.3

ISBN 978-7-5381-4994-4

Ⅰ.仓… Ⅱ.①小…②钟… Ⅲ.①办公室-建筑设计②仓库-改建 Ⅳ.TU243

中国版本图书馆CIP数据核字（2007）第028924号

出版发行：辽宁科学技术出版社
（地址：沈阳市和平区十一纬路25号 邮编：110003）
印 刷 者：利丰雅高印刷（深圳）有限公司
经 销 者：各地新华书店
幅面尺寸：255mm×255mm
印　　张：21
插　　页：4
字　　数：60千字
印　　数：1～3000
出版时间：2007年3月第1版
印刷时间：2007年3月第1次印刷
责任编辑：陈慈良
封面设计：静语轩
版式设计：静语轩
责任校对：徐 跃

定　　价：198.00元

联系电话：024-23284360
邮购热线：024-23284502
E-mail：lkzzb@mail.lnpge.com.cn
http://www.lnkj.vom.cn